1. Draw an array that shows 5 rows of 2.

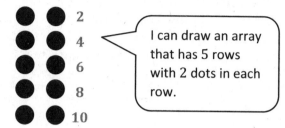

I can draw an array that has 5 rows with 2 dots in each row.

Write a multiplication sentence where the first factor represents the number of rows.

_____5_____ × _____2_____ = _____10_____

I can write a multiplication sentence with 5 as the first factor because 5 is the number of rows. The second factor is 2 because there are 2 dots in each row. I can skip-count by 2 to find the product, 10.

2. Draw an array that shows 2 rows of 5.

I can draw an array that has 2 rows with 5 dots in each row.

Write a multiplication sentence where the first factor represents the number of rows.

_____2_____ × _____5_____ = _____10_____

I can write a multiplication sentence with 2 as the first factor because 2 is the number of rows. The second factor is 5 because there are 5 dots in each row. I can skip-count by 5 to find the product, 10.

3. Why are the factors in your multiplication sentences in a different order?

The factors are in a different order because they mean different things. Problem 1 is 5 rows of 2, and Problem 2 is 2 rows of 5. In Problem 1, the 5 represents the number of rows. In Problem 2, the 5 represents the number of dots in each row.

The arrays show the commutative property. The order of the factors changed because the factors mean different things for each array. The product stayed the same for each array.

EUREKA MATH™

Lesson 7: Demonstrate the commutativity of multiplication, and practice related facts by skip-counting objects in array models.

©2018 Great Minds®. eureka-math.org

27

4. Write a multiplication sentence to match the number of groups. Skip-count to find the totals.

   a. 7 twos: $\underline{7 \times 2 = 14}$

   b. 2 sevens: $\underline{2 \times 7 = 14}$

> 7 twos is unit form. It means that there are 7 groups of 2. I can represent that with the multiplication equation $7 \times 2 = 14$. 2 sevens means 2 groups of 7, which I can represent with the multiplication equation $2 \times 7 = 14$.

> I see a pattern! 7 twos is equal to 2 sevens. It's the commutative property! The factors switched places and mean different things, but the product didn't change.

5. Find the unknown factor to make each equation true.

   $2 \times 8 = 8 \times \underline{2}$ $\underline{4} \times 2 = 2 \times 4$

> To make true equations, I need to make sure what's on the left of the equal sign is the same as (or equal to) what's on the right of the equal sign.

> I can use the commutative property to help me. I know that $2 \times 8 = 16$ and $8 \times 2 = 16$, so I can write 2 in the first blank. To solve the second problem, I know that $4 \times 2 = 8$ and $2 \times 4 = 8$. I can write 4 in the blank.

Lesson 7: Demonstrate the commutativity of multiplication, and practice related facts by skip-counting objects in array models.

EUREKA MATH™

Name _____Trisha_____   Date _____

1. a. Draw an array that shows 7 rows of 2.

   b. Write a multiplication sentence where the first factor represents the number of rows.

   ___7__ × __2__ = __14__

2. a. Draw an array that shows 2 rows of 7.

   b. Write a multiplication sentence where the first factor represents the number of rows.

   ___2__ × __7__ = __14__

3. a. Turn your paper to look at the arrays in Problems 1 and 2 in different ways. What is the same and what is different about them? They are same because they have same factors and product. They are diffrent because in #1 it was 7 rows of 2, in #2 it was 2 rows of 7. Then it swiched.

   b. Why are the factors in your multiplication sentences in a different order? Because the factor represents the number of groups so in #1 it was 7 rows of 2 so 7x2, but in #2 it was 2 rows of 7 so 2x7.

4. Write a multiplication sentence to match the number of groups. Skip-count to find the totals. The first one is done for you.

   a. 2 twos: __2 × 2 = 4__

   b. 3 twos: __3x2=6__

   c. 2 threes: __2x3=6__

   d. 2 fours: __2x4=8__

   e. 4 twos: __4x2=8__

   f. 5 twos: __5x2=10__

   g. 2 fives: __2x5=10__

   h. 6 twos: __6x2=12__

   i. 2 sixes: __2x6=12__

EUREKA MATH

Lesson 7: Demonstrate the commutativity of multiplication, and practice related facts by skip-counting objects in array models.

©2018 Great Minds®. eureka-math.org

29

5. Write and solve multiplication sentences where the second factor represents the size of the row.

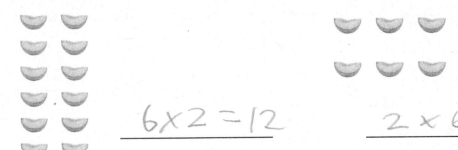

$6 \times 2 = 12$ $2 \times 6 = 12$

6. Angel writes $2 \times 8 = 8 \times 2$ in his notebook. Do you agree or disagree? Draw arrays to help explain your thinking.

I agree.

2×8

8×2

$8 \times 2 = 16$
$2 \times 8 = 16$

Yes, I agree.

7. Find the missing factor to make each equation true.

$2 \times 6 = 6 \times \underline{2}$ $\underline{7} \times 2 = 2 \times 7$ $9 \times 2 = \underline{2} \times 9$ $2 \times \underline{10} = 10 \times 2$

8. Tamia buys 2 bags of candy. Each bag has 7 pieces of candy in it.

    a. Draw an array to show how many pieces of candy Tamia has altogether.

    b. Write and solve a multiplication sentence to describe the array.

    $2 \times 7 = 14$

    c. Use the commutative property to write and solve a different multiplication sentence for the array.

    $2 \times 7 = 14$

    $7 \times 2 = 14$

Lesson 7:   Demonstrate the commutativity of multiplication, and practice related facts by skip-counting objects in array models.

©2018 Great Minds®. eureka-math.org

EUREKA MATH™

1. Find the unknowns that make the equations true. Then, draw a line to match related facts.

   a. $3 + 3 + 3 + 3 =$ _____ **12**

   b. $3 \times 7 =$ _____ **21**

   c. 5 threes + 1 three = _____ **6 threes**

   d. $3 \times 6 =$ _____ **18**

   e. _____ **12** $= 4 \times 3$

   f. $21 = 7 \times$ _____ **3**

   > $3 + 3 + 3 + 3$ is the same as 4 threes or 4×3, which equals 12. These equations are related because they both show that 4 groups of 3 equal 12.

   > 5 threes + 1 three = 6 threes. 6 threes is the same as 6 threes of 3 or 6×3, which equals 18. I can use the commutative property to match this equation with $3 \times 6 = 18$.

   > I can use the commutative property to match $3 \times 7 = 21$ and $21 = 7 \times 3$.

2. Fred puts 3 stickers on each page of his sticker album. He puts stickers on 7 pages.

   a. Use circles to draw an array that represents the total number of stickers in Fred's sticker album.

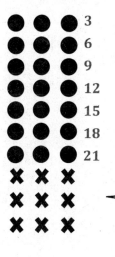

   3
   6
   9
   12
   15
   18
   21

   > I can draw an array with 7 rows to represent the 7 pages of the sticker album. I can draw 3 circles in each row to represent the 3 stickers that Fred puts on each page.

   > I can draw 3 more rows of 3 to represent the 3 pages and 3 stickers on each page that Fred adds to his sticker album in part (c).

b. Use your array to write and solve a multiplication sentence to find Fred's total number of stickers.

$7 \times 3 = 21$

Fred puts 21 stickers in his sticker album.

> I can write the multiplication equation $7 \times 3 = 21$ to find the total because there are 7 rows in my array with 3 circles in each row. I can use my array to skip-count to find the total, 21.

c. Fred adds 3 more pages to his sticker album. He puts 3 stickers on each new page. Draw x's to show the new stickers on the array in part (a).

d. Write and solve a multiplication sentence to find the new total number of stickers in Fred's sticker album.

$24, 27, 30$

$10 \times 3 = 30$

Fred has a total of 30 stickers in his sticker album.

> I can continue to skip-count by three from 21 to find the total, 30. I can write the multiplication equation $10 \times 3 = 30$ to find the total because there are 10 rows in my array with 3 in each row. The number of rows changed, but the size of each row stayed the same.

     Lesson 8:   Demonstrate the commutativity of multiplication, and practice related
                                             facts by skip-counting objects in array models.

EUREKA MATH

Name __Tvisha__

Date _____

1. Draw an array that shows 6 rows of 3.

2. Draw an array that shows 3 rows of 6.

3. Write multiplication expressions for the arrays in Problems 1 and 2. Let the first factor in each expression represent the number of rows. Use the commutative property to make sure the equation below is true.

$$\underline{\quad 6 \quad} \times \underline{\quad 3 \quad} = \underline{\quad 3 \quad} \times \underline{\quad 6 \quad}$$

Problem 1   **Problem 2**

4. Write a multiplication sentence for each expression. You might skip-count to find the totals. The first one is done for you.

a. 5 threes: __5 × 3 = 15__

b. 3 fives: __3 × 5 = 15__

c. 6 threes: __6 × 3 = 18__

d. 3 sixes: __3 × 6 = 18__

e. 7 threes: __7 × 3 = 21__

f. 3 sevens: __3 × 7 = 21__

g. 8 threes: __8 × 3 = 24__

h. 3 nines: __9 × 3 = 27__

i. 10 threes: __10 × 3 = 30__

5. Find the unknowns that make the equations true. Then, draw a line to match related facts.

a. 3 + 3 + 3 + 3 + 3 + 3 = __18__

b. 3 × 5 = __15__

c. 8 threes + 1 three = __27__

d. 3 × 9 = __27__

e. __18__ = 6 × 3

f. 15 = 5 × __3__

EUREKA MATH

Lesson 8:   Demonstrate the commutativity of multiplication, and practice related facts by skip-counting objects in array models.

©2018 Great Minds®. eureka-math.org

33

6. Fernando puts 3 pictures on each page of his photo album. He puts pictures on 8 pages.

a. Use circles to draw an array that represents the total number of pictures in Fernando's photo album.

b. Use your array to write and solve a multiplication sentence to find Fernando's total number of pictures.

$3 \times 8 = 24$

c. Fernando adds 2 more pages to his book. He puts 3 pictures on each new page. Draw x's to show the new pictures on the array in Part (a).

d. Write and solve a multiplication sentence to find the new total number of pictures in Fernando's album.

$5 \times 8 = 40$

7. Ivania recycles. She gets 3 cents for every can she recycles.

a. How much money does Ivania make if she recycles 4 cans?

$\underline{4} \times \underline{3} = \underline{12}$ cents

b. How much money does Ivania make if she recycles 7 cans?

$\underline{7} \times \underline{3} = \underline{22}$ cents

Lesson 8:   Demonstrate the commutativity of multiplication, and practice related facts by skip-counting objects in array models.

©2018 Great Minds®. eureka-math.org

EUREKA MATH™

1. Matt organizes his baseball cards into 3 rows of three. Jenna adds 2 more rows of 3 baseball cards. Complete the equations to describe the total number of baseball cards in the array.

a. $(3 + 3 + 3) + (3 + 3) =$ _____15_____

b. 3 threes + ____2____ threes = ____5____ threes

c. ____5____ × 3 = ____15____

> The multiplication equation for this array is 5 × 3 = 15 because there are 5 threes or 5 rows of 3, which is a total of 15 baseball cards.

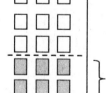

> The total for Matt's baseball cards (the unshaded rectangles) can be represented by 3 + 3 + 3 because there are 3 rows of 3 baseball cards. The total for Jenna's baseball cards (the shaded rectangles) can be represented by 3 + 3 because there are 2 rows of 3 baseball cards. This can be represented in unit form with 3 threes +2 threes, which equals 5 threes.

2. $8 × 3 =$ ____24____

> I can find the product of 8 × 3 using the array and the equations below. This problem is different than the problem above because now I am finding two products and subtracting instead of adding.

> The multiplication equation for the whole array is 10 × 3 = 30. The multiplication equation for the shaded part is 2 × 3 = 6.

$10 × 3 = \underline{30}$

$2 × 3 = \underline{6}$

$30 - \underline{6} = 24$

$\underline{8} × 3 = 24$

> To solve 8 × 3, I can think of 10 × 3 because that's an easier fact. I can subtract the product of 2 × 3 from the product of 10 × 3.
> 30 − 6 = 24, so 8 × 3 = 24.

EUREKA MATH™

Lesson 9: Find related multiplication facts by adding and subtracting equal groups in array models.

©2018 Great Minds®. eureka-math.org

35

Name _____ Date _____

1. Dan organizes his stickers into 3 rows of four. Irene adds 2 more rows of stickers. Complete the equations to describe the total number of stickers in the array.

a. $(4 + 4 + 4) + (4 + 4) =$ _____

b. 3 fours + _____ fours = _____ fours

c. _____ × 4 = _____

2. $7 × 2 =$ _____

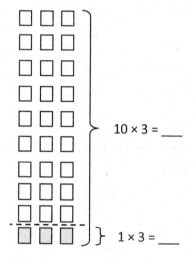

$6 × 2 =$ ___

$1 × 2 =$ ___

$12 + 2 =$ _____

_____ × 2 = 14

3. $9 × 3 =$ _____

$10 × 3 =$ ___

$1 × 3 =$ ___

$30 -$ _____ = 27

_____ × 3 = 27

Lesson 9: Find related multiplication facts by adding and subtracting equal groups in array models.

©2018 Great Minds®. eureka-math.org

37

4. Franklin collects stickers. He organizes his stickers in 5 rows of four.

    a. Draw an array to represent Franklin's stickers. Use an x to show each sticker.

    b. Solve the equation to find Franklin's total number of stickers. $5 \times 4 =$ _____

5. Franklin adds 2 more rows. Use circles to show his new stickers on the array in Problem 4(a).

    a. Write and solve an equation to represent the circles you added to the array.

                          _____ $\times 4 =$ _____

    b. Complete the equation to show how you add the totals of 2 multiplication facts to find Franklin's total number of stickers.

                          _____ $+$ _____ $= 28$

    c. Complete the unknown to show Franklin's total number of stickers.

                          _____ $\times 4 = 28$

Lesson 9:   Find related multiplication facts by adding and subtracting equal groups in array models.

©2018 Great Minds®. eureka-math.org

EUREKA
MATH™

1. Use the array to help you fill in the blanks.

   $6 \times 2 =$ ___**12**___

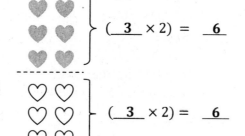

The dotted line in the array shows how I can break apart 6×2 into two smaller facts. Then I can add the products of the smaller facts to find the product of 6×2.

(**3** × 2) = **6**

(**3** × 2) = **6**

I know the first factor in each equation is 3 because there are 3 rows in each of the smaller arrays. The product for each array is 6.

$(3 \times 2) + (3 \times 2) =$ **6** + **6**

6 $\times 2 =$ **12**

The expressions in the parentheses represent the smaller arrays. I can add the products of these expressions to find the total number of hearts in the array. The products of the smaller expressions are both 6. $6 + 6 = 12$, so $6 \times 2 = 12$.

Hey, look! It's a doubles fact! $6 + 6 = 12$. I know my doubles facts, so this is easy to solve!

Lesson 10:   Model the distributive property with arrays to decompose units as a strategy to multiply.

©2018 Great Minds®. eureka-math.org

39

2. Lilly puts stickers on a piece of paper. She puts 3 stickers in each row.

   a. Fill in the equations to the right. Use them to draw arrays that show the stickers on the top and bottom parts of Lilly's paper.

I know there are 3 stickers in each row, and this equation also tells me that there are 12 stickers in all on the top of the paper. I can skip-count by 3 to figure out how many rows of stickers there. 3, 6, 9, 12. I skip-counted 4 threes, so there are 4 rows of 3 stickers. Now I can draw an array with 4 rows of 3.

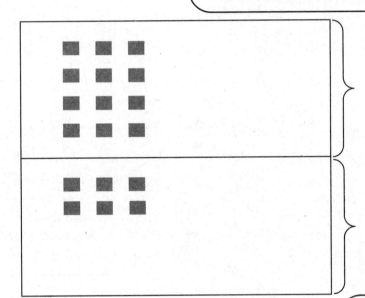

$\underline{\quad 4 \quad} \times 3 = 12$

$\underline{\quad 2 \quad} \times 3 = 6$

I see 6 rows of 3 altogether. I can use the products of these two smaller arrays to solve 6×3.

I can use the same strategy to find the number of rows in this equation. I skip-counted 2 threes, so there are 2 rows of 3 stickers. Now I can draw an array with 2 rows of 3.

Lesson 10: Model the distributive property with arrays to decompose units as a strategy to multiply.

©2018 Great Minds®. eureka-math.org

Name _____   Date _____

1. 6 × 3 = _____

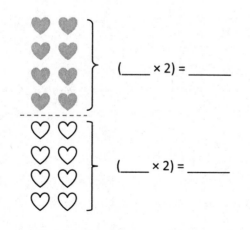

(4 × 3) = 12

(2 × 3) = _____

12 + _____ = _____

6 × 3 = _____

2. 8 × 2 = _____

(____ × 2) = _____

(____ × 2) = _____

(4 × 2) + (4 × 2) = _____ + _____

____ × 2 = _____

EUREKA MATH

Lesson 10:   Model the distributive property with arrays to decompose units as a strategy to multiply.

41

©2018 Great Minds®. eureka-math.org

3. Adriana organizes her books on shelves. She puts 3 books in each row.

    a. Fill in the equations on the right. Use them to draw arrays that show the books on Adriana's top and
        bottom shelves.

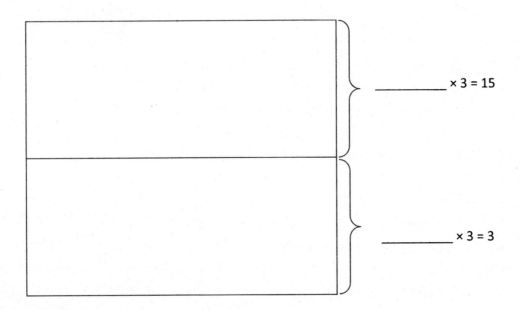

_____ × 3 = 15

_____ × 3 = 3

    b. Adriana calculates the total number of books as shown below. Use the array you drew to help explain
        Adriana's calculation.

$$6 \times 3 = 15 + 3 = 18$$

Lesson 10:   Model the distributive property with arrays to decompose units as a
                      strategy to multiply.

©2018 Great Minds®. eureka-math.org

EUREKA MATH™

1. Mr. Russell organizes 18 clipboards equally into 3 boxes. How many clipboards are in each box? Model the problem with both an array and a labeled tape diagram. Show each column as the number of clipboards in each box.

> I can draw an array with 3 columns because each column represents 1 box of clipboards. I can draw rows of 3 dots until I have a total of 18 dots. I can count how many dots are in each column to solve the problem.

> I know the total number of clipboards is 18, and there are 3 boxes of clipboards. I need to figure out how many clipboards are in each box. I can think of this as division, $18 \div 3 =$ ___, or as multiplication, $3 \times$ ___ $= 18$.

? clipboards

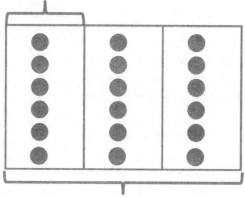

> I can draw 3 units in my tape diagram to represent the 3 boxes of clipboards. I can label the whole tape diagram with "18 clipboards". I can label one unit in the tape diagram with "? clipboards" because that's what I am solving for. I can draw 1 dot in each unit until I have a total of 18 dots.

18 *clipboards*

There are ___**6**___ clipboards in each box.

> Look, my array and tape diagram both show units of 6. The columns in my array each have 6 dots, and the units in my tape diagram each have a value of 6.

> I know the answer is 6 because my array has 6 dots in each column. My tape diagram also shows the answer because there are 6 dots in each unit.

EUREKA MATH™

Lesson 11: Model division as the unknown factor in multiplication using arrays and tape diagrams.

©2018 Great Minds®. eureka-math.org

43

2. Caden reads 2 pages in his book each day. How many days will it take him to read a total of 12 pages?

> This problem is different than the other problem because the known information is the total and the size of each group. I need to figure out how many groups there are.

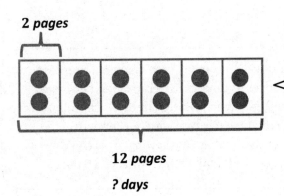

> I can draw an array where each column represents the number of pages Caden reads each day. I can keep drawing columns of 2 until I have a total of 12.

2 pages

> I can use my array to help me draw a tape diagram. I can draw 6 units of 2 in my tape diagram because my array shows 6 columns of 2.

12 pages

? days

12 ÷ 2 = 6

> I know the answer is 6 because my array shows 6 columns of 2, and my tape diagram shows 6 units of 2.

It will take Caden 6 days to read a total of 12 pages.

> I can write a statement to answer the question.

Lesson 11:   Model division as the unknown factor in multiplication using arrays and tape diagrams.

EUREKA MATH™

Name _____ Date _____

1. Fred has 10 pears. He puts 2 pears in each basket. How many baskets does he have?

   a. Draw an array where each column represents the number of pears in each basket.

   _____ ÷ 2 = _____

   b. Redraw the pears in each basket as a unit in the tape diagram. Label the diagram with known and unknown information from the problem.

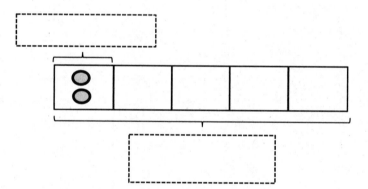

2. Ms. Meyer organizes 15 clipboards equally into 3 boxes. How many clipboards are in each box? Model the problem with both an array and a labeled tape diagram. Show each column as the number of clipboards in each box.

   There are _____ clipboards in each box.

EUREKA
MATH™

Lesson 11: Model division as the unknown factor in multiplication using arrays and tape diagrams.

©2018 Great Minds®. eureka-math.org

45

3. Sixteen action figures are arranged equally on 2 shelves. How many action figures are on each shelf? Model the problem with both an array and a labeled tape diagram. Show each column as the number of action figures on each shelf.

4. Jasmine puts 18 hats away. She puts an equal number of hats on 3 shelves. How many hats are on each shelf? Model the problem with both an array and a labeled tape diagram. Show each column as the number of hats on each shelf.

5. Corey checks out 2 books a week from the library. How many weeks will it take him to check out a total of 14 books?

Lesson 11:   Model division as the unknown factor in multiplication using arrays and tape diagrams.

©2018 Great Minds®. eureka-math.org

EUREKA
MATH™

1. Mrs. Harris divides 14 flowers equally into 7 groups for students to study. Draw flowers to find the number in each group. Label known and unknown information on the tape diagram to help you solve.

> I know the total number of flowers and the number of groups. I need to solve for the number of flowers in each group.

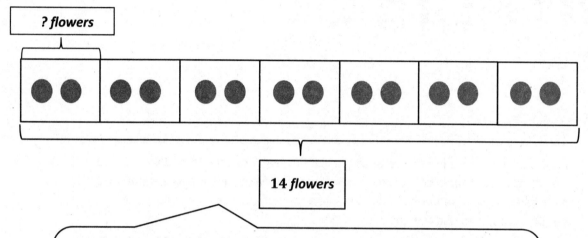

> **? flowers**

> **14 flowers**

> I can label the value of the tape diagram as "14 flowers". The number of units in the tape diagram, 7, represents the number of groups. I can label the unknown, which is the value of each unit, as "? flowers". I can draw 1 flower in each unit until I have a total of 14 flowers. I can draw dots instead of flowers to be more efficient!

> I can use my tape diagram to solve the problem by counting the number of dots in each unit.

$7 \times \underline{\quad 2 \quad} = 14$

$14 \div 7 = \underline{\quad 2 \quad}$

There are __2__ flowers in each group.

EUREKA MATH

Lesson 12: Interpret the quotient as the number of groups or the number of objects in each group using units of 2.

©2018 Great Minds®. eureka-math.org

47

2. Lauren finds 2 rocks each day for her rock collection. How many days will it take Lauren to find 16 rocks for her rock collection?

> I know the total is 16 rocks. I know Lauren finds 2 rocks each day, which is the size of each group. I need to figure out how many days it will take her to collect 16 rocks. The unknown is the number of groups.

2 rocks

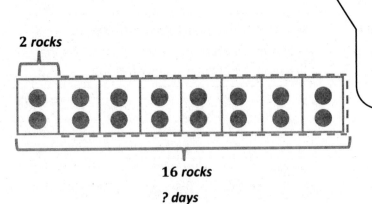

16 rocks

? days

> I can draw a tape diagram to solve this problem. I can draw a unit of 2 to represent the 2 rocks that Lauren collects each day. I can draw a dotted line to estimate the total days. I can draw units of 2 until I have a total of 16 rocks. I can count the number of units to find the answer.

$16 \div 2 = 8$

> I know the answer is 8 because my tape diagram shows 8 units of 2.

It will take Lauren 8 days to find 16 rocks.

> I can write a statement to answer the question.

Lesson 12:   Interpret the quotient as the number of groups or the number of objects in each group using units of 2.

©2018 Great Minds®. eureka-math.org

EUREKA MATH™

Name _____ Date _____

1. Ten people wait in line for the roller coaster. Two people sit in each car. Circle to find the total number of cars needed.

$10 \div 2 =$ _____

There are _____ cars needed.

2. Mr. Ramirez divides 12 frogs equally into 6 groups for students to study. Draw frogs to find the number in each group. Label known and unknown information on the tape diagram to help you solve.

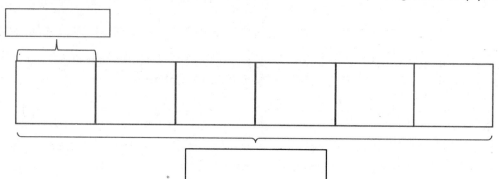

$6 \times$ _____ $= 12$

$12 \div 6 =$ _____

There are _____ frogs in each group.

3. Match.

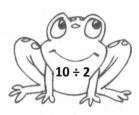 10 ÷ 2   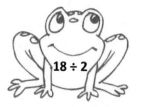 16 ÷ 2   18 ÷ 2    14 ÷ 2

 8   7    5   9

EUREKA MATH™

Lesson 12: Interpret the quotient as the number of groups or the number of objects in each group using units of 2.

49

©2018 Great Minds®. eureka-math.org

4. Betsy pours 16 cups of water to equally fill 2 bottles. How many cups of water are in each bottle? Label the tape diagram to represent the problem, including the unknown.

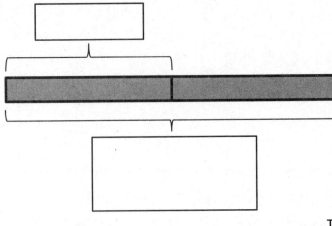

There are _____ cups of water in each bottle.

5. An earthworm tunnels 2 centimeters into the ground each day. The earthworm tunnels at about the same pace every day. How many days will it take the earthworm to tunnel 14 centimeters?

6. Sebastian and Teshawn go to the movies. The tickets cost $16 in total. The boys share the cost equally. How much does Teshawn pay?

Lesson 12:   Interpret the quotient as the number of groups or the number of objects in each group using units of 2.

©2018 Great Minds®. eureka-math.org

EUREKA
MATH™

1. Mr. Stroup's pet fish are shown below. He keeps 3 fish in each tank.

   a. Circle to show how many fish tanks he has. Then, skip-count to find the total number of fish.

   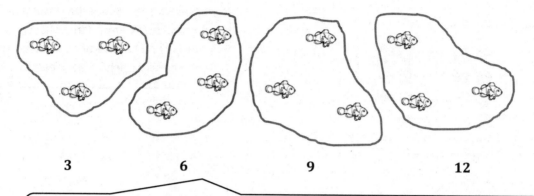

   3   6   9   12

   I can circle groups of 3 fish and skip-count by 3 to find the total number of fish. I can count the number of groups to figure out how many fish tanks Mr. Stroup has.

   Mr. Stroup has a total of 12 fish in 4 tanks.

   b. Draw and label a tape diagram to represent the problem.

   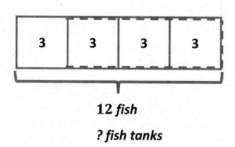

   12 fish

   ? fish tanks

   I can use the picture in part (a) to help me draw a tape diagram. Each fish tank has 3 fish, so I can label each unit with the number 3. I can draw a dotted line to estimate the total fish tanks. I can label the total as 12 fish. Then I can draw units of 3 until I have a total of 12 fish.

   The picture and the tape diagram both show that there are 4 fish tanks. The picture shows 4 equal groups of 3, and the tape diagram shows 4 units of 3.

   ___12___ ÷ 3 = ___4___

   Mr. Stroup has ___4___ fish tanks.

EUREKA MATH™

Lesson 13:   Interpret the quotient as the number of groups or the number of objects in each group using units of 3.

©2018 Great Minds®. eureka-math.org

51

2. A teacher has 21 pencils. They are divided equally among 3 students. How many pencils does each student get?

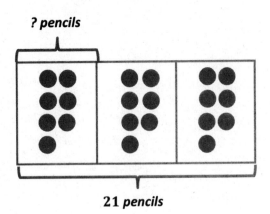

? pencils

21 pencils

> I can draw a tape diagram to solve this problem. I can draw 3 units to represent the 3 students. I can label the total number of pencils as 21 pencils. I need to figure out how many pencils each student gets.

> I know that I can divide 21 by 3 to solve. I don't know 21 ÷ 3, so I can draw one dot in each unit until I have a total of 21 dots. I can count the number of dots in one unit to find the quotient.

$21 ÷ 3 = 7$

> I know the answer is 7 because my tape diagram shows 3 units of 7.

Each student will get 7 pencils.

> I can write a statement to answer the question.

Lesson 13: Interpret the quotient as the number of groups or the number of objects in each group using units of 3.

©2018 Great Minds®. eureka-math.org

EUREKA MATH™

Name _____   Date _____

1. Fill in the blanks to make true number sentences.

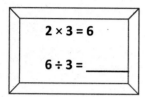
2 × 3 = 6

6 ÷ 3 = _____

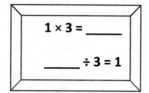
1 × 3 = _____

_____ ÷ 3 = 1

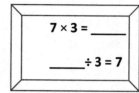
7 × 3 = _____

_____ ÷ 3 = 7

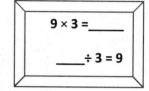
9 × 3 = _____

_____ ÷ 3 = 9

2. Ms. Gillette's pet fish are shown below. She keeps 3 fish in each tank.

   a. Circle to show how many fish tanks she has. Then, skip-count to find the total number of fish.

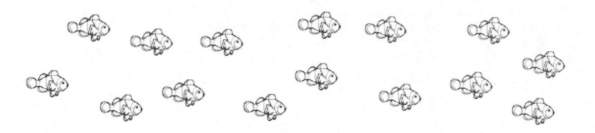

   b. Draw and label a tape diagram to represent the problem.

_____ ÷ 3 = _____

Ms. Gillette has _____ fish tanks.

EUREKA MATH™

Lesson 13:   Interpret the quotient as the number of groups or the number of
               objects in each group using units of 3.

©2018 Great Minds®. eureka-math.org

53

3. Juan buys 18 meters of wire. He cuts the wire into pieces that are each 3 meters long. How many pieces of wire does he cut?

4. A teacher has 24 pencils. They are divided equally among 3 students. How many pencils does each student get?

5. There are 27 third-graders working in groups of 3. How many groups of third-graders are there?

Lesson 13:   Interpret the quotient as the number of groups or the number of objects in each group using units of 3.

©2018 Great Minds®. eureka-math.org

EUREKA
MATH™

1. Mrs. Smith replaces 4 wheels on 3 cars. How many wheels does she replace? Draw and label a tape diagram to solve.

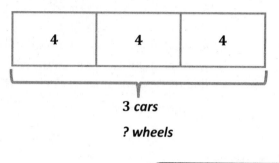

| 4 | 4 | 4 |

3 *cars*

? *wheels*

> I can draw a tape diagram with 3 units to represent the 3 cars. Each car has 4 wheels, so I can label each unit with the number 4. I need to find the total number of wheels.

4, 8, 12

$3 \times 4 = 12$

> I can skip-count by fours or multiply 3×4 to find how many wheels Mrs. Smith replaces.

Mrs. Smith replaces ___**12**___ wheels.

2. Thomas makes 4 necklaces. Each necklace has 7 beads. Draw and label a tape diagram to show the total number of beads Thomas uses.

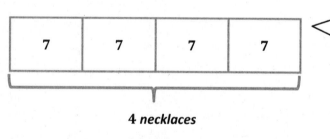

| 7 | 7 | 7 | 7 |

4 *necklaces*

? *beads*

> I can draw a tape diagram with 4 units to represent the 4 necklaces. I can label each unit in the tape diagram to show that every necklace has 7 beads. I need to find the total number of beads.

7, 14, 21, 28

4, 8, 12, 16, 20, 24, 28

$4 \times 7 = 28$

> I can skip-count 4 sevens, but sevens are still tricky for me. I can skip-count 7 fours instead! I can also multiply 4×7 to find how many beads Thomas uses.

Thomas uses ___**28**___ beads.

EUREKA MATH™

Lesson 14: Skip-count objects in models to build fluency with multiplication facts using units of 4.

55

©2018 Great Minds®. eureka-math.org

3. Find the total number of sides on 6 squares.

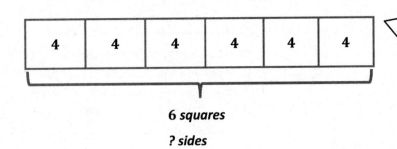

6 squares

? sides

I can draw a tape diagram with 6 units to represent the 6 squares. All squares have 4 sides, so I can label each unit with the number 4. I need to find the total number of sides.

4, 8, 12, 16, 20, 24

I can skip-count 6 fours or multiply 6 × 4 to find the total number of sides on 6 squares.

$6 \times 4 = 24$

There are 24 sides on 6 squares.

Lesson 14: Skip-count objects in models to build fluency with multiplication facts using units of 4.

©2018 Great Minds®. eureka-math.org

Name _____ Date _____

1. Skip-count by fours. Match each answer to the appropriate expression.

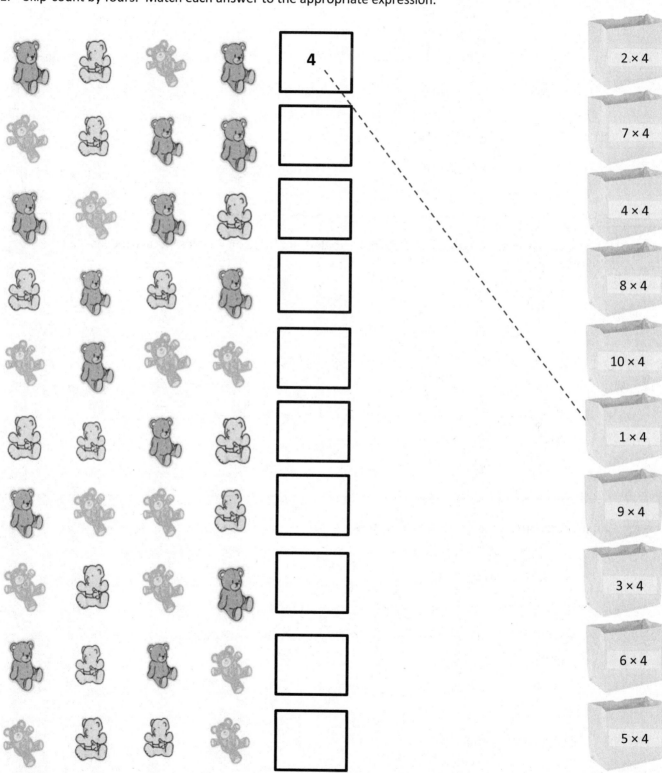

Lesson 14: Skip-count objects in models to build fluency with multiplication facts using units of 4.

©2018 Great Minds®. eureka-math.org

57

2. Lisa places 5 rows of 4 juice boxes in the refrigerator. Draw an array and skip-count to find the total number of juice boxes.

There are _____ juice boxes in total.

3. Six folders are placed on each table. How many folders are there on 4 tables? Draw and label a tape diagram to solve.

4. Find the total number of corners on 8 squares.

Lesson 14:   Skip-count objects in models to build fluency with multiplication facts using units of 4.

©2018 Great Minds®. eureka-math.org

EUREKA MATH™

1. Label the tape diagrams, and complete the equations. Then, draw an array to represent the problems.

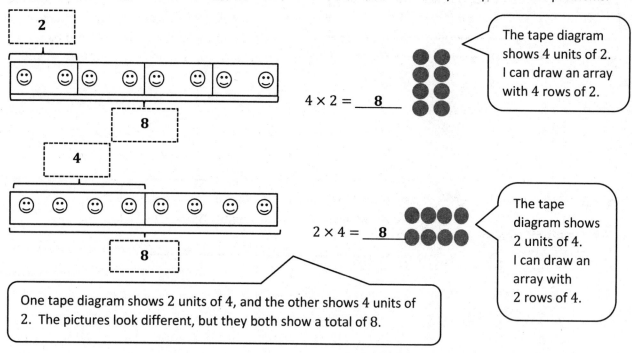

$4 \times 2 =$ ___**8**___

> The tape diagram shows 4 units of 2. I can draw an array with 4 rows of 2.

$2 \times 4 =$ ___**8**___

> The tape diagram shows 2 units of 4. I can draw an array with 2 rows of 4.

> One tape diagram shows 2 units of 4, and the other shows 4 units of 2. The pictures look different, but they both show a total of 8.

2. 8 books cost $4 each. Draw and label a tape diagram to show the total cost of the books.

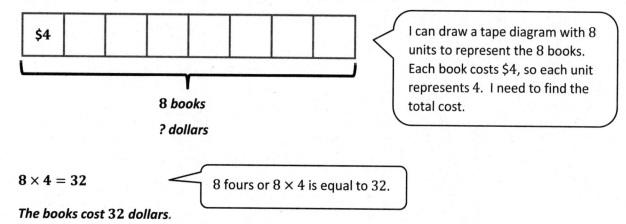

8 books

? dollars

> I can draw a tape diagram with 8 units to represent the 8 books. Each book costs $4, so each unit represents 4. I need to find the total cost.

8 × 4 = 32

> 8 fours or 8 × 4 is equal to 32.

The books cost 32 dollars.

EUREKA MATH

Lesson 15:   Relate arrays to tape diagrams to model the commutative property of multiplication.

©2018 Great Minds®. eureka-math.org

59

3. Liana reads 8 pages from her book each day. How many pages does Liana read in 4 days?

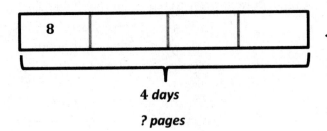

4 days

? pages

> I can draw a tape diagram with 4 units to represent the 4 days. Liana reads 8 pages each day, so each unit represents 8. I need to find the total number of pages.

$4 \times 8 = 32$

Liana reads 32 pages.

> I just solved 8×4, and I know that $8 \times 4 = 4 \times 8$. If 8 fours is equal to 32, then 4 eights is also equal to 32.

Lesson 15:   Relate arrays to tape diagrams to model the commutative property of multiplication.

EUREKA MATH™

Name _____ Date _____

1. Label the tape diagrams and complete the equations. Then, draw an array to represent the problems.

a.

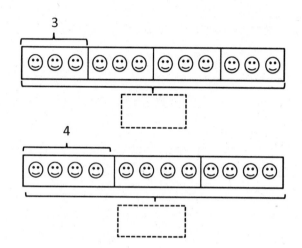

$4 \times 3 =$ _____

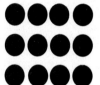

$3 \times 4 =$ _____

b.

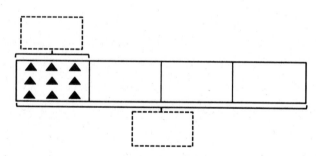

$4 \times$ _____ $=$ _____

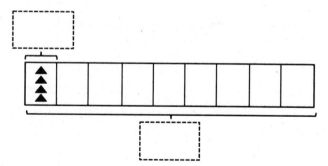

_____ $\times 4 =$ _____

EUREKA
MATH™

Lesson 15:   Relate arrays to tape diagrams to model the commutative property of multiplication.

©2018 Great Minds®. eureka-math.org

61

c.

_____ × 4 = _____

4 × _____ = _____

---

2. Seven clowns hold 4 balloons each at the fair. Draw and label a tape diagram to show the total number of balloons the clowns hold.

---

3. George swims 7 laps in the pool each day. How many laps does George swim after 4 days?

Lesson 15:   Relate arrays to tape diagrams to model the commutative property of multiplication.

©2018 Great Minds®. eureka-math.org

EUREKA
MATH

1. Label the array. Then, fill in the blanks below to make true number sentences.

$8 \times 3 = \underline{\ \ 24\ \ }$

$(5 \times 3) = \underline{\ \ 15\ \ }$

> I know that I can break apart 8 threes into 5 threes and 3 threes. I can add the products for 5×3 and 3×3 to find the product for 8×3.

$(3 \times 3) = \underline{\ \ 9\ \ }$

$$8 \times 3 = (5 \times 3) + (3 \times 3)$$
$$= \underline{\ \ 15\ \ } + \underline{\ \ 9\ \ }$$
$$= \underline{\ \ 24\ \ }$$

2. The array below shows one strategy for solving 8×4. Explain the strategy using your own words.

$(5 \times 4) = \underline{\ \ 20\ \ }$

> 8×4 is a tricky fact for me to solve, but 5×4 and 3×4 are both pretty easy facts. I can use them to help me!

$(3 \times 4) = \underline{\ \ 12\ \ }$

I split apart the 8 rows of 4 into 5 rows of 4 and 3 rows of 4. I split the array there because my fives facts and my threes facts are easier than my eights facts. I know that $5 \times 4 = 20$ and $3 \times 4 = 12$. I can add those products to find that $8 \times 4 = 32$.

EUREKA MATH™

Lesson 16: Use the distributive property as a strategy to find related multiplication facts.

©2018 Great Minds®. eureka-math.org

63

Name _____Tvisha_____ Date _____

1. Label the array. Then, fill in the blanks below to make true number sentences.

a. **6 × 4 = _____**

$(5 \times 4) =$ _20_

$(\underline{1} \times 4) =$ _4_   **(6 × 4)** = (5 × 4) + (____ × 4)

= _20_ + _____

= _____

b. **8 × 4 = _____**

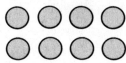

$(5 \times 4) =$ _____

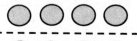

$(\underline{\quad} \times 4) =$ _____

(8 × 4) = (5 × 4) + (____ × 4)

= _____ + _____

= _____

EUREKA MATH

Lesson 16: Use the distributive property as a strategy to find related multiplication facts.

©2018 Great Minds®. eureka-math.org

65

2. Match the multiplication expressions with their answers.

4 × 6   4 × 7   4 × 8   4 × 9

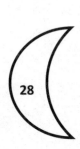  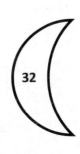 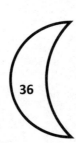 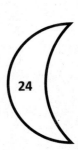

28   32   36   24

---

3. The array below shows one strategy for solving 9 × 4. Explain the strategy using your own words.

(5 × 4) = _____

(4 × 4) = _____

Lesson 16:   Use the distributive property as a strategy to find related multiplication facts.

©2018 Great Minds®. eureka-math.org

EUREKA
MATH

1. The baker packs 20 muffins into boxes of 4. Draw and label a tape diagram to find the number of boxes she packs.

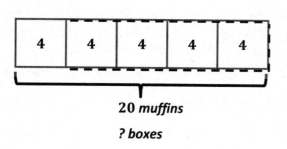

20 *muffins*

? *boxes*

$20 \div 4 =$ ___5___

The baker packs 5 boxes.

> I can draw a tape diagram. Each box has 4 muffins, so I can draw a unit and label it 4. I can draw a dotted line to estimate the total number of boxes, because I don't yet know how many boxes there are. I do know the total, so I'll label that as 20 muffins. I'll solve by drawing units of 4 in the dotted part of my tape diagram until I have a total of 20 muffins. Then I can count the number of units to see how many boxes of muffins the baker packs.

2. The waiter arranges 12 plates into 4 equal rows. How many plates are in each row?

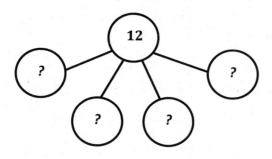

> I can use a number bond to solve. I know that the total number of plates is 12 and that the 12 plates are in 4 rows. Each part in the number bond represents a row of plates.

$12 \div 4 =$ ___3___

$3 \times 4 =$ ___12___

There are 3 plates in each row.

> I can divide to solve. I can also think of this as multiplication with an unknown factor.

3. A teacher has 20 erasers. She divides them equally between 4 students. She finds 12 more erasers and divides these equally between the 4 students as well. How many erasers does each student receive?

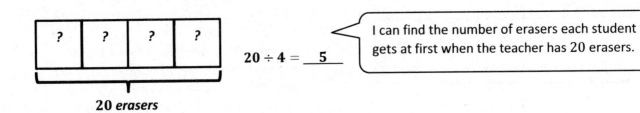

$20 \div 4 = \underline{\quad 5 \quad}$

> I can find the number of erasers each student gets at first when the teacher has 20 erasers.

20 erasers

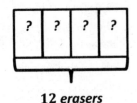

$12 \div 4 = \underline{\quad 3 \quad}$

> I can find how many erasers each student gets when the teacher finds 12 more erasers.

12 erasers

5 erasers + 3 erasers = __8__ erasers.

> I can add to find how many total erasers each student gets.

Each student receives 8 erasers.

Lesson 17: Model the relationship between multiplication and division.

©2018 Great Minds®. eureka-math.org

EUREKA MATH™

Name _____   Date _____

1. Use the array to complete the related equations.

$1 \times 4 =$ _____   _____ $\div 4 = 1$

$2 \times 4 =$ _____   _____ $\div 4 = 2$

_____ $\times 4 = 12$ $12 \div 4 =$ _____

_____ $\times 4 = 16$ $16 \div 4 =$ _____

_____ $\times$ _____ $= 20$ $20 \div$ _____ $=$ _____

_____ $\times$ _____ $= 24$ $24 \div$ _____ $=$ _____

_____ $\times 4 =$ _____   _____ $\div 4 =$ _____

_____ $\times 4 =$ _____   _____ $\div 4 =$ _____

_____ $\times$ _____ $=$ _____   _____ $\div$ _____ $=$ _____

_____ $\times$ _____ $=$ _____   _____ $\div$ _____ $=$ _____

2. The teacher puts 32 students into groups of 4. How many groups does she make? Draw and label a tape diagram to solve.

3. The store clerk arranges 24 toothbrushes into 4 equal rows. How many toothbrushes are in each row?

4. An art teacher has 40 paintbrushes. She divides them equally among her 4 students. She finds 8 more brushes and divides these equally among the students, as well. How many brushes does each student receive?

Lesson 17: Model the relationship between multiplication and division.

EUREKA
MATH™

1. Match the number bond on an apple with the equation on a bucket that shows the same total.

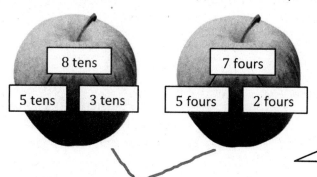

8 tens

5 tens | 3 tens

7 fours

5 fours | 2 fours

The number bonds in the apples help me see how I can find the total by adding the two smaller parts together. I can match the apples with the equations below that show the same two parts and total.

$(5 \times 4) + (2 \times 4) = 28$

$(5 \times 10) + (3 \times 10) = 80$

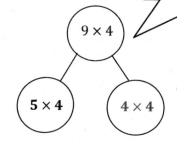

2. Solve.

$9 \times 4 = \underline{\ 36\ }$

I can think of this total as 9 fours. There are many ways to break apart 9 fours, but I'm going to break it apart as 5 fours and 4 fours because 5 is a friendly number.

9 × 4

5 × 4   4 × 4

I can use the number bond to help me fill in the blanks. Adding the **products** of these two smaller facts helps me find the product of the larger fact.

$(\underline{\ 5\ } \times 4) + (\underline{\ 4\ } \times 4) = 9 \times 4$

$\underline{\ 20\ } + \underline{\ 16\ } = \underline{\ 36\ }$

$9 \times 4 = \underline{\ 36\ }$

3. Mia solves 7×3 using the break apart and distribute strategy. Show an example of what Mia's work might look like below.

5 threes $+ 2$ threes $= 7$ threes

$(5 \times 3) + (2 \times 3) = 7 \times 3$

$15 + 6 = 21$

I can use the number bond to help me write the equations. Then I can find the products of the two smaller facts and add them to find the product of the larger fact.

The number bond helps me see the break apart and distribute strategy easily. I can think of 7×3 as 7 threes. Then I can break it apart as 5 threes and 2 threes.

Lesson 18:   Apply the distributive property to decompose units.

EUREKA MATH

Name _____ Date _____

1. Match.

7 tens

5 tens   2 tens

8 fours

5 fours   3 fours

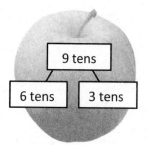

9 tens

6 tens   3 tens

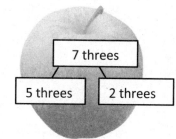

7 threes

5 threes   2 threes

(5 × 4) + (3 × 4) = 32

(5 × 3) + (2 × 3) = 21

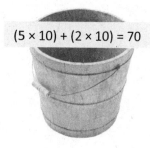

(5 × 10) + (2 × 10) = 70

(6 × 10) + (3 × 10) = 90

2. 9 × 4 = _____

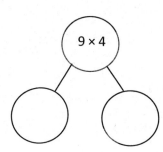

9 × 4

(_____ × 4) + (_____ × 4) = 9 × 4

_____ + _____ = _____

9 × 4 = _____

3. Lydia makes 10 pancakes. She tops each pancake with 4 blueberries. How many blueberries does Lydia use in all? Use the break apart and distribute strategy, and draw a number bond to solve.

Lydia uses _____ blueberries in all.

4. Steven solves 7 × 3 using the break apart and distribute strategy. Show an example of what Steven's work might look like below.

5. There are 7 days in 1 week. How many days are there in 10 weeks?

Lesson 18:   Apply the distributive property to decompose units.

©2018 Great Minds®. eureka-math.org

EUREKA
MATH

1. Solve.

   $28 \div 4 =$ __7__

   △ △ △ △

   △ △ △ △

   △ △ △ △ $(20 \div 4) =$ __5__

   △ △ △ △

   △ △ △ △

   - - - - - - - - - - - -

   △ △ △ △

   △ △ △ △ $(8 \div 4) =$ __2__

   > $(28 \div 4) = (20 \div 4) + ($ __8__ $\div 4)$
   >
   > $=$ __5__ $+$ __2__
   >
   > $=$ __7__

   This shows how we can add the quotients of two smaller facts to find the quotient of the larger one. The array can help me fill in the blanks.

   This array shows a total of 28 triangles. I see that the dotted line breaks apart the array after the fifth row. There are 5 fours above the dotted line and 2 fours below the dotted line.

Match equal expressions.

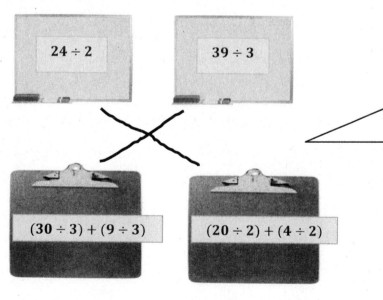

$24 \div 2$

$39 \div 3$

$(30 \div 3) + (9 \div 3)$

$(20 \div 2) + (4 \div 2)$

I can match the larger division problem found on the whiteboard to the two smaller division problems added together on the clipboard below.

EUREKA MATH

2. Chloe draws the array below to find the answer to $48 \div 4$. Explain Chloe's strategy.

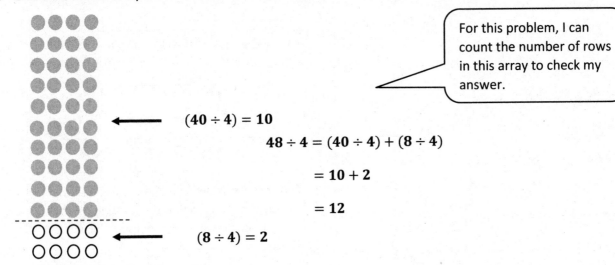

For this problem, I can count the number of rows in this array to check my answer.

$(40 \div 4) = 10$

$$48 \div 4 = (40 \div 4) + (8 \div 4)$$

$$= 10 + 2$$

$$= 12$$

$(8 \div 4) = 2$

Chloe breaks apart 48 as 10 fours and 2 fours. 10 fours equals 40, and 2 fours equals 8. So, she does 40 ÷ 4 and 8 ÷ 4 and adds the answers to get 48 ÷ 4, which equals 12.

EUREKA
MATH™

Name _____ Date _____

1. Label the array. Then, fill in the blanks to make true number sentences.

a. 18 ÷ 3 = _____

△ △ △
△ △ △   (9 ÷ 3) = 3
△ △ △
- - - - - - - - - - - - -
△ △ △
△ △ △   (9 ÷ 3) = _____
△ △ △

```
(18 ÷ 3) = (9 ÷ 3) + (9 ÷ 3)

      =  __3__ + _____

      =  __6__
```

b. 21 ÷ 3 = _____

△ △ △
△ △ △
△ △ △   (15 ÷ 3) = 5
△ △ △
△ △ △
- - - - - - - - - - - - -
△ △ △
△ △ △   (6 ÷ 3) = _____

```
(21 ÷ 3) = (15 ÷ 3) + (6 ÷ 3)

      =  __5__ + _____

      =  _____
```

c. 24 ÷ 4 = _____

△ △ △ △
△ △ △ △
△ △ △ △   (20 ÷ 4) = _____
△ △ △ △
△ △ △ △
- - - - - - - - - - - - - - -
△ △ △ △   (4 ÷ 4) = _____

```
(24 ÷ 4) = (20 ÷ 4) + (____ ÷ 4)

      =  _____ + _____

      =  _____
```

d. 36 ÷ 4 = _____

△ △ △ △
△ △ △ △
△ △ △ △
△ △ △ △   (20 ÷ 4) = _____
△ △ △ △
- - - - - - - - - - - - - - -
△ △ △ △
△ △ △ △
△ △ △ △   (16 ÷ 4) = _____
△ △ △ △

```
(36 ÷ 4) = (____ ÷ 4) + (____ ÷ 4)

      =  _____ + _____

      =  _____
```

2. Match equal expressions.

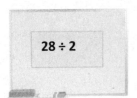   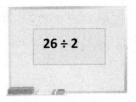

$28 \div 2$ $33 \div 3$ $36 \div 3$ $26 \div 2$

 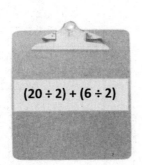 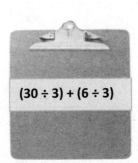 

$(30 \div 3) + (3 \div 3)$ $(20 \div 2) + (6 \div 2)$ $(30 \div 3) + (6 \div 3)$ $(20 \div 2) + (8 \div 2)$

3. Alex draws the array below to find the answer to $35 \div 5$. Explain Alex's strategy.

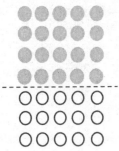

Lesson 19:   Apply the distributive property to decompose units.

©2018 Great Minds®. eureka-math.org

EUREKA
MATH™

1. Thirty-five students are eating lunch at 5 tables. Each table has the same number of students.

   a. How many students are sitting at each table?

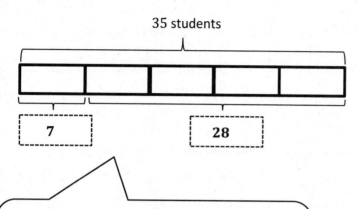

35 students

7

28

I know there are a total of 35 students eating lunch at 5 tables. I know each table has the same number of students. I need to figure out how many students are sitting at each table. The unknown is the size of each group.

Each unit in my tape diagram represents 1 table. Since there are 35 students and 5 tables, I can divide 35 by 5 to find that each table has 7 students. This tape diagram shows that there are 5 units of 7 for a total of 35.

$35 \div 5 = 7$

There are 7 students sitting at each table.

   b. How many students are sitting at 4 tables?

$4 \times 7 = 28$

There are 28 students sitting at 4 tables.

I can write a number sentence and a statement to answer the question.

Since I now know there are 7 students sitting at each table, I can multiply the number of tables, 4, by 7 to find that there are 28 students sitting at 4 tables. I can see this in the tape diagram: 4 units of 7 equal 28.

Lesson 20:   Solve two-step word problems involving multiplication and division, and assess the reasonableness of answers.

©2018 Great Minds®. eureka-math.org

2. The store has 30 notebooks in packs of 3. Six packs of notebooks are sold. How many packs of notebooks are left?

> I know the total is 30 notebooks. I know the notebooks are in packs of 3. First I need to figure out how many total packs of notebooks are in the store.

> I can draw a tape diagram that shows 30 notebooks in packs of 3. I can find the total number of packs by dividing 30 by 3 to get 10 total packs of notebooks.

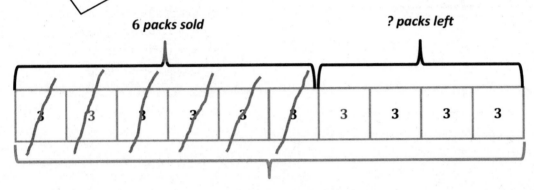

6 packs sold   **? packs left**

30 notebooks

? total packs

> Now that I know the total number of packs is 10, I can find the number of packs that are left.

$30 \div 3 = 10$

There are a total of 10 packs of notebooks at the store.

$10 - 6 = 4$

There are 4 packs of notebooks left.

> I can show the packs that were sold on my tape diagram by crossing off 6 units of 3. Four units of 3 are not crossed off, so there are 4 packs of notebooks left. I can write a subtraction equation to represent the work on my tape diagram.

Lesson 20:   Solve two-step word problems involving multiplication and division, and assess the reasonableness of answers.

EUREKA MATH™

Name _____   Date _____

1. Jerry buys a pack of pencils that costs $3. David buys 4 sets of markers. Each set of markers also costs $3.

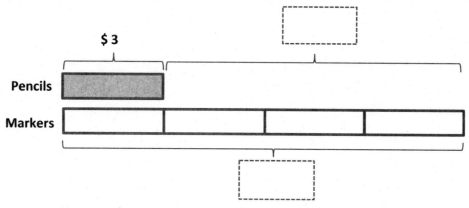

a. What is the total cost of the markers?

b. How much more does David spend on 4 sets of markers than Jerry spends on a pack of pencils?

2. Thirty students are eating lunch at 5 tables. Each table has the same number of students.

a. How many students are sitting at each table?

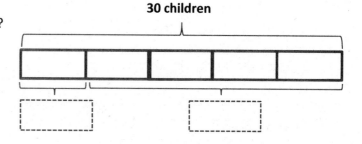

b. How many students are sitting at 4 tables?

EUREKA MATH™

Lesson 20:   Solve two-step word problems involving multiplication and division, and assess the reasonableness of answers.

©2018 Great Minds®. eureka-math.org

81

3. The teacher has 12 green stickers and 15 purple stickers. Three students are given an equal number of each color sticker. How many green and purple stickers does each student get?

---

4. Three friends go apple picking. They pick 13 apples on Saturday and 14 apples on Sunday. They share the apples equally. How many apples does each person get?

---

5. The store has 28 notebooks in packs of 4. Three packs of notebooks are sold. How many packs of notebooks are left?

Lesson 20:   Solve two-step word problems involving multiplication and division, and assess the reasonableness of answers.

©2018 Great Minds®. eureka-math.org

EUREKA
MATH™

1. John has a reading goal. He checks out 3 boxes of 7 books from the library. After finishing them, he realizes that he beat his goal by 5 books! Label the tape diagrams to find John's reading goal.

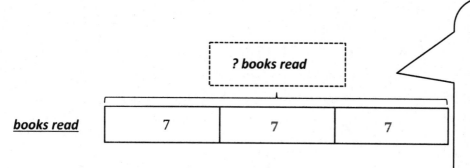

Each unit in this tape diagram represents 1 box of John's library books. The number of books in each box (the size) is 7 books. So I can multiply 3×7 to find the number of books John reads.

$3 \times 7 = 21$

John reads **21** *books.*

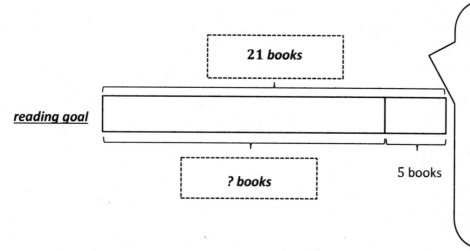

I can draw a tape diagram that shows 21 as the total because John reads 21 books. I can label one part as 5 because John beat his reading goal by 5 books. When I know a total and one part, I know I can subtract to find the other part.

$21 - 5 = 16$

John's goal was to read ___16___ books.

I can check back to see if my statement answers the question.

EUREKA MATH

Lesson 21: Solve two-step word problems involving all four operations, and assess the reasonableness of answers.

©2018 Great Minds®. eureka-math.org

83

2. Mr. Kim plants 20 trees around the neighborhood pond. He plants equal numbers of maple, pine, spruce, and birch trees. He waters the spruce and birch trees before it gets dark. How many trees does Mr. Kim still need to water? Draw and label a tape diagram.

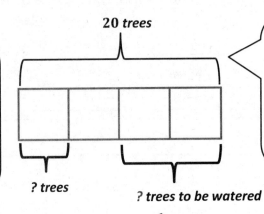

20 trees

I know Mr. Kim plants a total of 20 trees. He plants an equal number of 4 types of trees. This is the number of groups. So, the unknown is the size of each group.

I can draw a tape diagram that has 4 units to represent the 4 types of trees. I can label the whole as 20, and I can divide 20 by 4 to find the value of each unit.

? trees

? trees to be watered

I know that Mr. Kim waters the spruce and birch trees, so he still needs to water the maple and pine trees. I can see from my tape diagram that 2 units of 5 trees still need to be watered. I can multiply 2 × 5 to find that 10 trees still need to be watered.

$20 \div 4 = 5$
Mr. Kim plants 5 of each type of tree.

$2 \times 5 = 10$
Mr. Kim still needs to water 10 trees.

$20 - 10 = 10$
Mr. Kim still needs to water 10 trees.

Or I can subtract the number of trees watered, 10, from the total number of trees to find the answer.

Lesson 21:   Solve two-step word problems involving all four operations, and assess the reasonableness of answers.

©2018 Great Minds®. eureka-math.org

EUREKA MATH™

Name _____   Date _____

1. Tina eats 8 crackers for a snack each day at school. On Friday, she drops 3 and only eats 5. Write and solve an equation to show the total number of crackers Tina eats during the week.

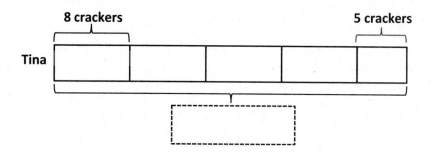

Tina eats _____ crackers.

_____

2. Ballio has a reading goal. He checks 3 boxes of 9 books out from the library. After finishing them, he realizes that he beat his goal by 4 books! Label the tape diagrams to find Ballio's reading goal.

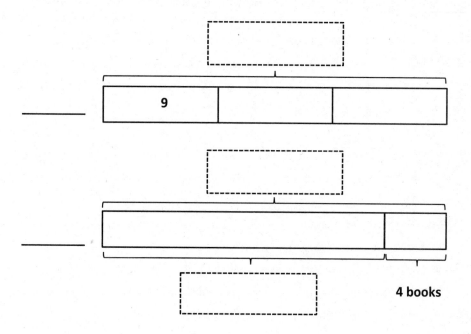

Ballio's goal is to read _____ books.

EUREKA MATH™

Lesson 21:   Solve two-step word problems involving all four operations, and assess the reasonableness of answers.

©2018 Great Minds®. eureka-math.org

85

3. Mr. Nguyen plants 24 trees around the neighborhood pond. He plants equal numbers of maple, pine, spruce, and birch trees. He waters the spruce and birch trees before it gets dark. How many trees does Mr. Nguyen still need to water? Draw and label a tape diagram.

4. Anna buys 24 seeds and plants 3 in each pot. She has 5 pots. How many more pots does Anna need to plant all of her seeds?

Lesson 21: Solve two-step word problems involving all four operations, and assess the reasonableness of answers.

©2018 Great Minds®. eureka-math.org

EUREKA MATH

Grade 3
Module 2

Eric	19 seconds
Woo	20 seconds
Sharon	24 seconds
Steven	18 seconds
Joyce	22 seconds

The table to the right shows how much time it takes each of the 5 students to run 100 meters.

a. Who is the fastest runner?

Steven is the fastest runner.

> I know Steven is the fastest runner because the chart shows me that he ran 100 meters in the least number of seconds, 18 seconds.

b. Who is the slowest runner?

Sharon is the slowest runner.

> I know Sharon is the slowest runner because the chart shows me that she ran 100 meters in the most number of seconds, 24 seconds.

c. How many seconds faster did Eric run than Sharon?

$24 - 19 = 5$

Eric ran 5 seconds faster than Sharon.

> I can subtract Eric's time from Sharon's time to find how much faster Eric ran than Sharon. I can use the compensation strategy to think of subtracting $24 - 19$ as $25 - 20$ to get 5. It is much easier for me to subtract $25 - 20$ than $24 - 19$.

EUREKA MATH

Lesson 1: Explore time as a continuous measurement using a stopwatch.

89

©2018 Great Minds®. eureka-math.org

Follow the directions to label the number line below.

a. Susan practices piano between 3:00 p.m. and 4:00 p.m. Label the first and last tick marks as 3:00 p.m. and 4:00 p.m.

3:00 p.m.   **4:00 p.m.**

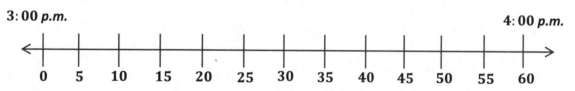

> I can label this first tick mark as 3:00 p.m. and the last tick mark as 4:00 p.m. to show the hour interval Susan practices piano.

b. Each interval represents 5 minutes. Count by fives starting at 0, or 3:00 p.m. Label each 5-minute interval below the number line up to 4:00 p.m.

3:00 p.m.   **4:00 p.m.**

0   5   10   15   20   25   30   35   40   45   50   55   60

> I know there are 60 minutes between 3:00 p.m. and 4:00 p.m. I can label 0 minutes below where I wrote 3:00 p.m. and label 60 minutes below where I wrote 4:00 p.m.

> I can skip-count by fives to label each 5-minute interval from left to right, starting with 0 and ending with 60.

EUREKA
MATH™

Lesson 2:   Relate skip-counting by fives on the clock and telling time to a
             continuous measurement model, the number line.

©2018 Great Minds®. eureka-math.org

93

c. Susan warms up her fingers by playing the scales until 3:10 p.m. Plot a point on the number line to represent this time. Above the point, write *W*.

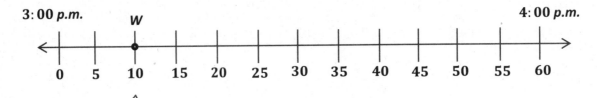

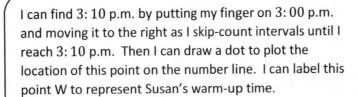

I can find 3:10 p.m. by putting my finger on 3:00 p.m. and moving it to the right as I skip-count intervals until I reach 3:10 p.m. Then I can draw a dot to plot the location of this point on the number line. I can label this point W to represent Susan's warm-up time.

Lesson 2: Relate skip-counting by fives on the clock and telling time to a continuous measurement model, the number line.

EUREKA MATH™

Name _____   Date _____

Follow the directions to label the number line below.

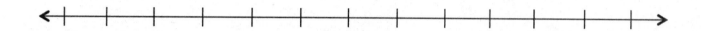

a. The basketball team practices between 4:00 p.m. and 5:00 p.m. Label the first and last tick marks as 4:00 p.m. and 5:00 p.m.

b. Each interval represents 5 minutes. Count by fives starting at 0, or 4:00 p.m. Label each 5-minute interval below the number line up to 5:00 p.m.

c. The team warms up at 4:05 p.m. Plot a point on the number line to represent this time. Above the point, write W.

d. The team shoots free throws at 4:15 p.m. Plot a point on the number line to represent this time. Above the point, write F.

e. The team plays a practice game at 4:25 p.m. Plot a point on the number line to represent this time. Above the point, write G.

f. The team has a water break at 4:50 p.m. Plot a point on the number line to represent this time. Above the point, write B.

g. The team reviews their plays at 4:55 p.m. Plot a point on the number line to represent this time. Above the point, write P.

Lesson 2:   Relate skip-counting by fives on the clock and telling time to a continuous measurement model, the number line.

©2018 Great Minds®. eureka-math.org

95

The clock shows what time Caleb starts playing outside on Monday afternoon.

a. What time does he start playing outside?

Caleb starts playing outside at 2:32 p.m.

I can find the minutes on this analog clock by counting by fives and ones, beginning on the 12, as zero minutes.

Start

b. He plays outside for 19 minutes. What time does he finish playing?

Caleb finishes playing outside at 2:51 p.m.

I can use different strategies to find the time Caleb finishes playing. The most efficient strategy is to add 20 minutes to 2:32 to get 2:52, and then subtract 1 minute to get 2:51.

c. Draw hands on the clock to the right to show what time Caleb finishes playing.

Finish

I can check my answer from part (b) by counting by fives and ones on the clock, and then draw the hands on the clock. My minute hand is exactly at 51 minutes, but my hour hand is close to the 3 since it is almost 3:00.

EUREKA MATH

Lesson 3: Count by fives and ones on the number line as a strategy to tell time to the nearest minute on the clock.

97

©2018 Great Minds®. eureka-math.org

d.   Label the first and last tick marks with 2:00 p.m. and 3:00 p.m. Then, plot Caleb's start and finish times. Label his start time with a *B* and his finish time with an *F*.

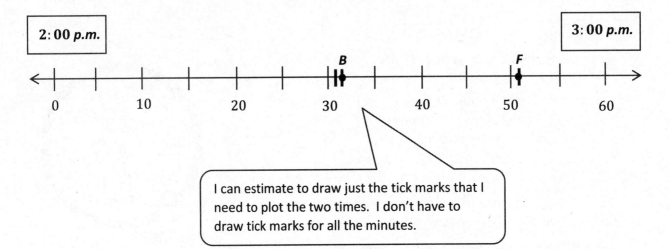

I can estimate to draw just the tick marks that I need to plot the two times. I don't have to draw tick marks for all the minutes.

Count by fives and ones on the number line as a strategy to tell time to the nearest minute on the clock.

©2018 Great Minds®. eureka-math.org

EUREKA
MATH™

Name _____ Date _____

1. Plot points on the number line for each time shown on a clock below. Then, draw lines to match the clocks to the points.

4:00 p.m.   5:00 p.m.

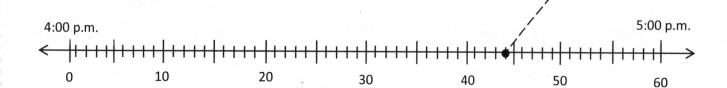

0   10   20   30   40   50   60

---

2. Julie eats dinner at 6:07 p.m. Draw hands on the clock below to show what time Julie eats dinner.

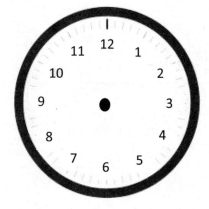

3. P.E. starts at 1:32 p.m. Draw hands on the clock below to show what time P.E. starts.

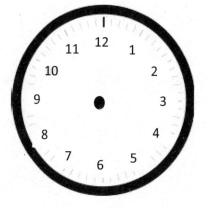

EUREKA MATH

Lesson 3: Count by fives and ones on the number line as a strategy to tell time to the nearest minute on the clock.

99

©2018 Great Minds®. eureka-math.org

4. The clock shows what time Zachary starts playing with his action figures.

   a.   What time does he start playing with his action figures?

Start

   b.   He plays with his action figures for 23 minutes.
         What time does he finish playing?

   c.   Draw hands on the clock to the right to show what time
         Zachary finishes playing.

Finish

   d.   Label the first and last tick marks with 2:00 p.m. and 3:00 p.m. Then, plot Zachary's start and finish
         times. Label his start time with a *B* and his finish time with an *F*.

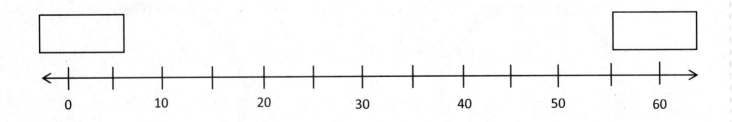

Lesson 3:   Count by fives and ones on the number line as a strategy to tell time to
                   the nearest minute on the clock.

©2018 Great Minds®. eureka-math.org

EUREKA
MATH™

Use a number line to answer the problems below.

1. Celina cleans her room for 42 minutes. She starts at 9:04 a.m. What time does Celina finish cleaning her room?

> I can draw a number line to help me figure out when Celina finishes cleaning her room. On the number line, I can label the first tick mark 0 and the last tick mark 60. Then I can label the hours and the 5-minute intervals.

9:00 a.m.   **10:00 a.m.**

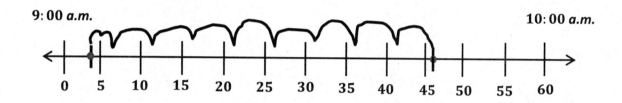

0   5   10   15   20   25   30   35   40   45   50   55   60

Celina finishes cleaning her room at 9:46 a.m.

> I can plot 9:04 a.m. on the number line. Then I can count 2 minutes to 9:06 and 40 minutes by fives until 9:46. 42 minutes after 9:04 a.m. is 9:46 a.m.

2. The school orchestra puts on a concert for the school. The concert lasts 35 minutes. It ends at 1:58 p.m. What time did the concert start?

1:00 p.m.   **2:00 p.m.**

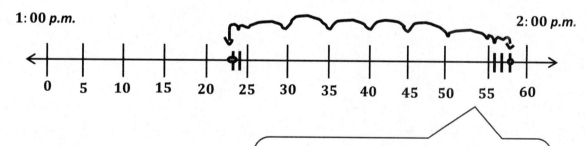

0   5   10   15   20   25   30   35   40   45   50   55   60

The concert started at 1:23 p.m.

> I can plot 1:58 p.m. on the number line. Then I can count backwards from 1:58 by ones to 1:55, by fives to 1:25, and by ones to 1:23. 1:23 p.m. is 35 minutes before 1:58 p.m.

EUREKA
MATH™

Lesson 4:   Solve word problems involving time intervals within 1 hour by counting backward and forward using the number line and clock.

©2018 Great Minds®. eureka-math.org

101

Name _____   Date _____

Record your homework start time on the clock in Problem 6.

Use a number line to answer Problems 1 through 4.

1. Joy's mom begins walking at 4:12 p.m. She stops at 4:43 p.m. How many minutes does she walk?

Joy's mom walks for _____ minutes.

_____

2. Cassie finishes softball practice at 3:52 p.m. after practicing for 30 minutes. What time did Cassie's practice start?

Cassie's practice started at _____ p.m.

_____

3. Jordie builds a model from 9:14 a.m. to 9:47 a.m. How many minutes does Jordie spend building his model?

Jordie builds for _____ minutes.

_____

4. Cara finishes reading at 2:57 p.m. She reads for a total of 46 minutes. What time did Cara start reading?

Cara started reading at _____ p.m.

Lesson 4:   Solve word problems involving time intervals within 1 hour by counting backward and forward using the number line and clock.

©2018 Great Minds®. eureka-math.org

103

5. Jenna and her mom take the bus to the mall. The clocks below show when they leave their house and when they arrive at the mall. How many minutes does it take them to get to the mall?

Time when they leave home:

Time when they arrive at the mall:

6. Record your homework start time:

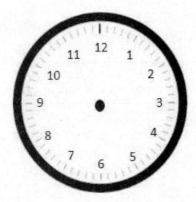

Record the time when you finish Problems 1–5:

How many minutes did you work on Problems 1–5?

Lesson 4: Solve word problems involving time intervals within 1 hour by counting backward and forward using the number line and clock.

EUREKA MATH™

Luke exercises. He stretches for 8 minutes, runs for 17 minutes, and walks for 10 minutes.

a. How many total minutes does he spend exercising?

I can draw a tape diagram to show all the known information. I see all the parts are given, but the whole is unknown. So, I can label the whole with a question mark.

? minutes

| 8 min | 17 min | 10 min |

I can estimate to draw the parts of my tape diagram to match the lengths of the minutes. 8 minutes is the shortest time, so I can draw it as the shortest unit. 17 minutes is the longest time, so I can draw it as the longest unit.

$8 + 17 + 10 = 35$

Luke spends a total of 35 *minutes exercising.*

I can write an addition equation to find the total number of minutes Luke spends exercising. I also need to remember to write a statement that answers the question.

EUREKA MATH™

Lesson 5: Solve word problems involving time intervals within 1 hour by adding and subtracting on the number line.

©2018 Great Minds®. eureka-math.org

105

b. Luke wants to watch a movie that starts at 1:55 p.m. It takes him 10 minutes to take a shower and 15 minutes to drive to the theater. If Luke starts exercising at 1:00 p.m., can he make it on time for the movie? Explain your reasoning.

> I can draw a number line to show my reasoning. I can plot the starting time as 1:35 because I know it takes Luke 35 minutes to exercise from part (a). Then I can add 10 minutes for his shower and an additional 15 minutes for the drive to the theater.

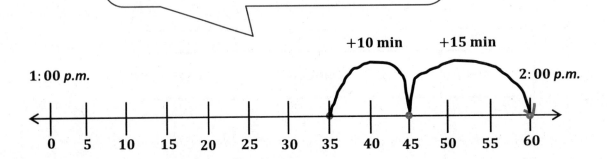

No, Luke can't make it on time for the movie. From the number line, I can see that he will be five minutes late.

> I can see on the number line that Luke will be at the theater at 2:00 p.m. The movie starts at 1:55 p.m., so he'll be 5 minutes too late.

106   **Lesson 5:**   Solve word problems involving time intervals within 1 hour by adding
                                   and subtracting on the number line.

                        ©2018 Great Minds®. eureka-math.org

EUREKA
MATH™

Name _____   Date _____

1. Abby spent 22 minutes working on her science project yesterday and 34 minutes working on it today. How many minutes did Abby spend working on her science project altogether? Model the problem on the number line, and write an equation to solve.

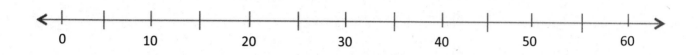

Abby spent _____ minutes working on her science project.

2. Susanna spends a total of 47 minutes working on her project. How many more minutes than Susanna does Abby spend working? Draw a number line to model the problem, and write an equation to solve.

3. Peter practices violin for a total of 55 minutes over the weekend. He practices 25 minutes on Saturday. How many minutes does he practice on Sunday?

EUREKA
MATH™

Lesson 5:   Solve word problems involving time intervals within 1 hour by adding and subtracting on the number line.

©2018 Great Minds®. eureka-math.org

107

4. a. Marcus gardens. He pulls weeds for 18 minutes, waters for 13 minutes, and plants for 16 minutes. How many total minutes does he spend gardening?

b. Marcus wants to watch a movie that starts at 2:55 p.m. It takes 10 minutes to drive to the theater. If Marcus starts the yard work at 2:00 p.m., can he make it on time for the movie? Explain your reasoning.

5. Arelli takes a short nap after school. As she falls asleep, the clock reads 3:03 p.m. She wakes up at the time shown below. How long is Arelli's nap?

Lesson 5: Solve word problems involving time intervals within 1 hour by adding and subtracting on the number line.

©2018 Great Minds®. eureka-math.org

EUREKA MATH

1. Use the chart to help you answer the following questions:

1 kilogram	100 grams	10 grams	1 gram

a. Bethany puts a marker that weighs 10 grams on a pan balance. How many 1-gram weights does she need to balance the scale?

Bethany needs ten 1-gram weights to balance the scale.

> I know that it takes ten 1-gram weights to equal 10 grams.

b. Next, Bethany puts a 100-gram bag of beans on a pan balance. How many 10-gram weights does she need to balance the scale?

Bethany needs ten 10-gram weights to balance the scale.

> I know that it takes ten 10-gram weights to equal 100 grams.

c. Bethany then puts a book that weighs 1 kilogram on a pan balance. How many 100-gram weights does she need to balance the scale?

Bethany needs ten 100-gram weights to balance the scale.

> I know that it takes ten 100-gram weights to equal 1 kilogram, or 1,000 grams.

d. What pattern do you notice in parts (a)–(c)?

I notice that to make a weight in the chart it takes ten of the lighter weight to the right in the chart. For example, to make 100 grams, it takes ten 10-gram weights, and to make 1 kilogram, or 1,000 grams, it takes ten 100-gram weights. It's just like the place value chart!

Lesson 6:   Build and decompose a kilogram to reason about the size and weight of 1 kilogram, 100 grams, 10 grams, and 1 gram.   **109**

©2018 Great Minds®. eureka-math.org

2.   Read each digital scale. Write each weight using the word *kilogram* or *gram* for each measurement.

_____153 *grams*_____

_____3 *kilograms*_____

I can write 153 grams because I know that the letter g is used to abbreviate grams.

I can write 3 kilograms because I know that the letters kg are used to abbreviate kilograms.

Lesson 6:   Build and decompose a kilogram to reason about the size and weight of 1 kilogram, 100 grams, 10 grams, and 1 gram.

EUREKA MATH™

Name _____ Date _____

1. Use the chart to help you answer the following questions:

1 kilogram	100 grams	10 grams	1 gram

a. Isaiah puts a 10-gram weight on a pan balance. How many 1-gram weights does he need to balance the scale?

b. Next, Isaiah puts a 100-gram weight on a pan balance. How many 10-gram weights does he need to balance the scale?

c. Isaiah then puts a kilogram weight on a pan balance. How many 100-gram weights does he need to balance the scale?

d. What pattern do you notice in Parts (a–c)?

EUREKA
MATH™

Lesson 6: Build and decompose a kilogram to reason about the size and weight
of 1 kilogram, 100 grams, 10 grams, and 1 gram.

©2018 Great Minds®. eureka-math.org

111

2. Read each digital scale. Write each weight using the word *kilogram* or *gram* for each measurement.

3 kg

6 kg

450 g

_____

_____

_____

907 g

11 kg

1 kg

_____

_____

_____

Lesson 6: Build and decompose a kilogram to reason about the size and weight of 1 kilogram, 100 grams, 10 grams, and 1 gram.

©2018 Great Minds®. eureka-math.org

EUREKA MATH™

1. Match each object with its approximate weight.

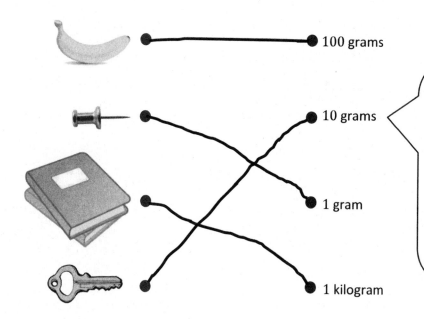

100 grams

10 grams

1 gram

1 kilogram

> I know that the tack is the lightest object, so it must weigh about 1 gram. I also know that the books are the heaviest, so they must weigh about 1 kilogram. I know that the key is lighter than the banana, so the key must weigh about 10 grams and the banana must weigh about 100 grams.

2. Jessica weighs her dog on a digital scale. She writes 8, but she forgets to record the unit. Which unit of measurement is correct, grams or kilograms? How do you know?

   The weight of Jessica's dog needs to be recorded as 8 kilograms. Kilograms is the correct unit because 8 grams is about the same weight as 8 paperclips. It wouldn't make sense for her dog to weigh about the same as 8 paperclips.

3. Read and write the weight below. Write the word *kilogram* or *gram* with the measurement.

> I know the unit is grams because there is a letter g on the scale. I can use the image to the right of the scale to determine that each tick mark between 140 grams and 150 grams represents 1 gram. The fruit weighs 146 grams.

   146 grams

EUREKA MATH

Lesson 7: Develop estimation strategies by reasoning about the weight in kilograms of a series of familiar objects to establish mental benchmark measures.

©2018 Great Minds®. eureka-math.org

113

Name _____   Date _____

1. Match each object with its approximate weight.

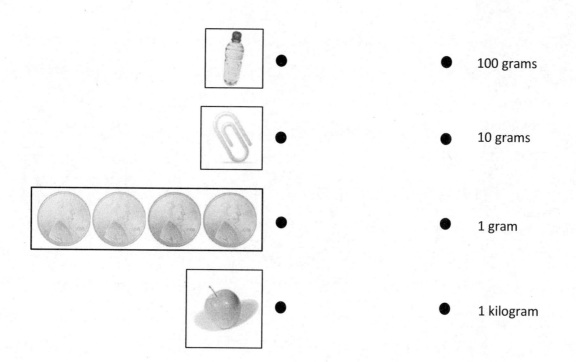

2. Alicia and Jeremy weigh a cell phone on a digital scale. They write down 113 but forget to record the unit. Which unit of measurement is correct, grams or kilograms? How do you know?

EUREKA MATH™

Lesson 7:   Develop estimation strategies by reasoning about the weight in kilograms of a series of familiar objects to establish mental benchmark measures.

©2018 Great Minds®. eureka-math.org

115

3. Read and write the weights below. Write the word *kilogram* or *gram* with the measurement.

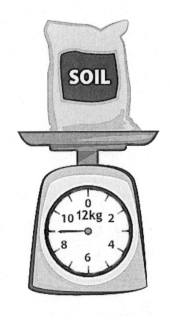

_____

_____

_____

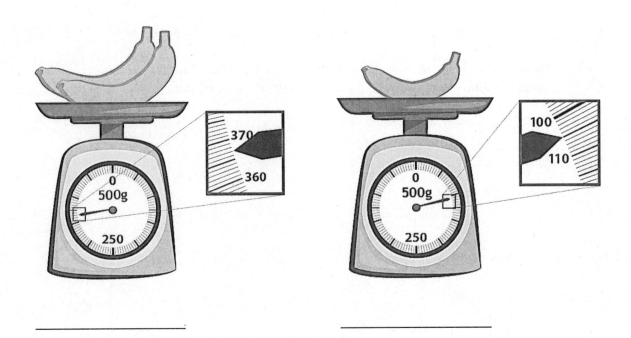

_____

_____

Lesson 7: Develop estimation strategies by reasoning about the weight in kilograms of a series of familiar objects to establish mental benchmark measures.

©2018 Great Minds®. eureka-math.org

EUREKA MATH™

The weights below show the weight of the apples in each bucket.

Bucket A
9 kg

Bucket B
7 kg

Bucket C
14 kg

> Bucket C weighs 14 kg, and Bucket B weighs 7 kg. I know that $14 - 7 = 7$, so Bucket C weighs 7 kg more.

a. The apples in Bucket __C__ are the heaviest.

b. The apples in Bucket __B__ are the lightest.

c. The apples in Bucket C are __7__ kilograms heavier than the apples in Bucket B.

d. What is the total weight of the apples in all three buckets?

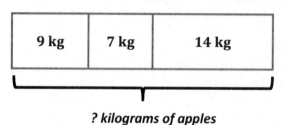

? kilograms of apples

$$9 + 7 + 14 = 30$$

The total weight of the apples is 30 kilograms.

> I can use a tape diagram to show the weight of each bucket of apples. Then, I can add each apple's weight to find the total weight of the apples.

e. Rebecca and her 2 sisters equally share all of the apples in Bucket A. How many kilograms of apples do they each get?

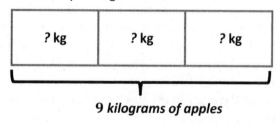

9 kilograms of apples

$$9 \div 3 = 3$$

Each sister gets 3 kilograms of apples.

> I know that I'm dividing 9 kilograms into 3 equal groups because 3 people are sharing the apples in Bucket A. When I know the total and the number of equal groups, I divide to find the size of each group!

EUREKA MATH   Lesson 8:   Solve one-step word problems involving metric weights within 100 and estimate to reason about solutions.   117

©2018 Great Minds®. eureka-math.org

f. Mason gives 3 kilograms of apples from Bucket B to his friend. He uses 2 kilograms of apples from Bucket B to make apple pies. How many kilograms of apples are left in Bucket B?

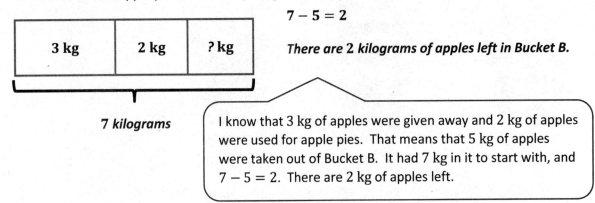

$7 - 5 = 2$

There are 2 kilograms of apples left in Bucket B.

7 kilograms

I know that 3 kg of apples were given away and 2 kg of apples were used for apple pies. That means that 5 kg of apples were taken out of Bucket B. It had 7 kg in it to start with, and $7 - 5 = 2$. There are 2 kg of apples left.

g. Angela picks another bucket of apples, Bucket D. The apples in Bucket C are 6 kilograms heavier than the apples in Bucket D. How many kilograms of apples are in Bucket D?

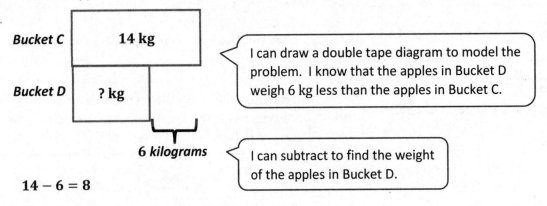

I can draw a double tape diagram to model the problem. I know that the apples in Bucket D weigh 6 kg less than the apples in Bucket C.

6 kilograms

I can subtract to find the weight of the apples in Bucket D.

$14 - 6 = 8$

There are 8 kilograms of apples in Bucket D.

h. What is the total weight of the apples in Buckets C and D?

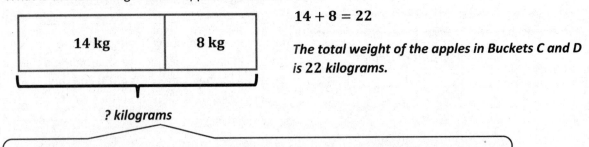

$14 + 8 = 22$

The total weight of the apples in Buckets C and D is 22 kilograms.

? kilograms

To find the total weight of the apples in Buckets C and D, I need to add. I know that $14 + 8 = 22$, so the total weight of the apples in Buckets C and D is 22 kilograms.

Lesson 8:   Solve one-step word problems involving metric weights within 100 and estimate to reason about solutions.

©2018 Great Minds®. eureka-math.org

EUREKA MATH™

Name _____ Date _____

1. The weights of 3 fruit baskets are shown below.

Basket A   Basket B   Basket C
12 kg   8 kg   16 kg

a. Basket _____ is the heaviest.

b. Basket _____ is the lightest.

c. Basket A is _____ kilograms heavier than Basket B.

d. What is the total weight of all three baskets?

2. Each journal weighs about 280 grams. What is total weight of 3 journals?

3. Ms. Rios buys 453 grams of strawberries. She has 23 grams left after making smoothies. How many grams of strawberries did she use?

Lesson 8: Solve one-step word problems involving metric weights within 100 and estimate to reason about solutions.

©2018 Great Minds®. eureka-math.org

119

4. Andrea's dad is 57 kilograms heavier than Andrea. Andrea weighs 34 kilograms.

   a.   How much does Andrea's dad weigh?

   b.   How much do Andrea and her dad weigh in total?

5. Jennifer's grandmother buys carrots at the farm stand. She and her 3 grandchildren equally share the carrots. The total weight of the carrots she buys is shown below.

   a.   How many kilograms of carrots will Jennifer get?

   b.   Jennifer uses 2 kilograms of carrots to bake muffins. How many kilograms of carrots does she have left?

Lesson 8:   Solve one-step word problems involving metric weights within 100 and estimate to reason about solutions.

EUREKA MATH™

©2018 Great Minds®. eureka-math.org

1. Ben makes 4 batches of cookies for the bake sale. He uses 5 milliliters of vanilla for each batch. How many milliliters of vanilla does he use in all?

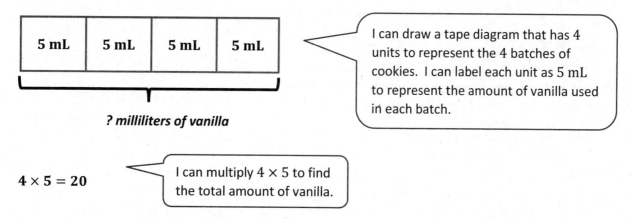

I can draw a tape diagram that has 4 units to represent the 4 batches of cookies. I can label each unit as 5 mL to represent the amount of vanilla used in each batch.

$4 \times 5 = 20$

I can multiply 4×5 to find the total amount of vanilla.

Ben uses 20 milliliters of vanilla.

2. Mrs. Gillette pours 3 glasses of juice for her children. Each glass holds 321 milliliters of juice. How much juice does Mrs. Gillette pour in all?

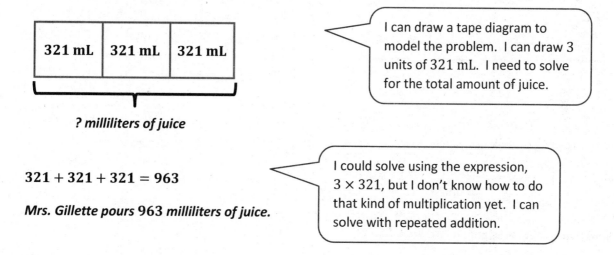

I can draw a tape diagram to model the problem. I can draw 3 units of 321 mL. I need to solve for the total amount of juice.

$321 + 321 + 321 = 963$

Mrs. Gillette pours 963 milliliters of juice.

I could solve using the expression, 3×321, but I don't know how to do that kind of multiplication yet. I can solve with repeated addition.

EUREKA MATH

Lesson 9: Decompose a liter to reason about the size of 1 liter, 100 milliliters, 10 milliliters, and 1 milliliter.

©2018 Great Minds®. eureka-math.org

121

3. Gabby uses a 4-liter bucket to give her pony water. How many buckets of water will Gabby need in order to give her pony 28 liters of water?

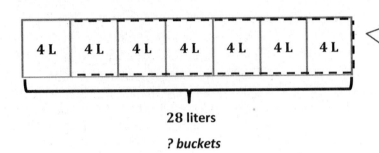

| 4 L | 4 L | 4 L | 4 L | 4 L | 4 L | 4 L |

28 liters

? buckets

I can draw a tape diagram. I know the total is 28 liters and the size of each unit is 4 liters. I need to solve for the number of units (buckets).

$28 \div 4 = 7$

Gabby needs 7 buckets of water.

Since I know the total and the size of each unit, I can divide to solve.

4. Elijah makes 12 liters of punch for his birthday party. He pours the punch equally into 4 bowls. How many liters of punch are in each bowl?

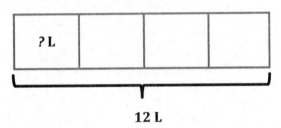

| ? L | | | |

12 L

I can draw a tape diagram. I know the total is 12 liters and there are 4 bowls or units. I need to solve for the number of liters in each bowl.

$12 \div 4 = 3$

Since I know the total and the number of units, I can divide to solve.

Elijah pours 3 liters of punch into each bowl.

I can divide to solve Problems 3 and 4, but the unknowns in each problem are different. In Problem 3, I solved for the number of groups/units. In Problem 4, I solved for the size of each group/unit.

Lesson 9:   Decompose a liter to reason about the size of 1 liter, 100 milliliters,
                                           10 milliliters, and 1 milliliter.

©2018 Great Minds®. eureka-math.org

EUREKA MATH™

Name _____ Date _____

1. Find containers at home that have a capacity of about 1 liter. Use the labels on containers to help you identify them.

   a.

Name of Container
Example: Carton of orange juice

   b. Sketch the containers. How do their sizes and shapes compare?

   _____

2. The doctor prescribes Mrs. Larson 5 milliliters of medicine each day for 3 days. How many milliliters of medicine will she take altogether?

EUREKA
MATH™

Lesson 9:   Decompose a liter to reason about the size of 1 liter, 100 milliliters, 10 milliliters, and 1 milliliter.

©2018 Great Minds®. eureka-math.org

123

3. Mrs. Goldstein pours 3 juice boxes into a bowl to make punch. Each juice box holds 236 milliliters. How much juice does Mrs. Goldstein pour into the bowl?

---

4. Daniel's fish tank holds 24 liters of water. He uses a 4-liter bucket to fill the tank. How many buckets of water are needed to fill the tank?

---

5. Sheila buys 15 liters of paint to paint her house. She pours the paint equally into 3 buckets. How many liters of paint are in each bucket?

Lesson 9:   Decompose a liter to reason about the size of 1 liter, 100 milliliters, 10 milliliters, and 1 milliliter.

EUREKA MATH

1. Estimate the amount of liquid in each container to the nearest liter.

The liquid in this container is between 3 liters and 4 liters. Since it is more than halfway to the next liter, 4 liters, I can estimate that there are about 4 liters of liquid.

__4 liters__

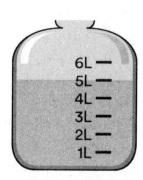

The liquid in this container is at exactly 5 liters.

__5 liters__

The liquid in this container is between 3 liters and 4 liters. Since it is less than halfway to the next liter, 4 liters, I can estimate that there are about 3 liters of liquid.

__3 liters__

EUREKA MATH™

Lesson 10:   Estimate and measure liquid volume in liters and milliliters using the vertical number line.

125

©2018 Great Minds®. eureka-math.org

2. Manny is comparing the capacity of buckets that he uses to water his vegetable garden. Use the chart to answer the questions.

Bucket	Capacity in Liters
Bucket 1	17
Bucket 2	12
Bucket 3	23

a. Label the number line to show the capacity of each bucket. Bucket 2 has been done for you.

I can use the tick marks to help me locate the correct place on the number line for each bucket. I can label Bucket 1 at 17 liters and Bucket 3 at 23 liters.

b. Which bucket has the greatest capacity?

Bucket 3 has the greatest capacity.

c. Which bucket has the smallest capacity?

Bucket 2 has the smallest capacity.

I can use the vertical number line to help me answer both of these questions. I can see that the point I plotted for Bucket 3 is higher up the number line than the others, so it has a larger capacity than the others. I also see that the point I plotted for Bucket 2 is lowest on the number line, so it has the smallest capacity.

d. Which bucket has a capacity of about 10 liters?

Bucket 2 has a capacity of about 10 liters.

I notice that Bucket 2 is closest to 10 liters, so it has a capacity of about 10 liters.

e. Use the number line to find how many more liters Bucket 3 holds than Bucket 2.

Bucket 3 holds 11 more liters than Bucket 2.

To solve this problem, I can count up on the number line from Bucket 2 to Bucket 3. I'll start at 12 liters because that is the capacity of Bucket 2. I count up 8 tick marks to 20 liters, and then I count 3 more tick marks to 23, which is the capacity of Bucket 3. I know that $8 + 3 = 11$, so Bucket 3 holds 11 more liters than Bucket 2.

Lesson 10: Estimate and measure liquid volume in liters and milliliters using the vertical number line.

©2018 Great Minds®. eureka-math.org

EUREKA MATH

Name_____ Date _____

1. How much liquid is in each container?

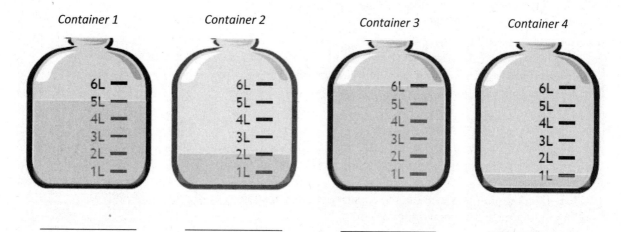

Container 1   Container 2   Container 3   Container 4

_____   _____   _____   _____

2. Jon pours the contents of Container 1 and Container 3 above into an empty bucket. How much liquid is in the bucket after he pours the liquid?

3. Estimate the amount of liquid in each container to the nearest liter.

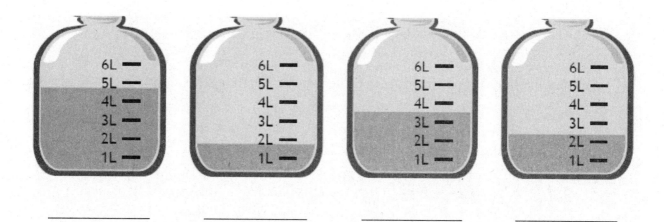

_____   _____   _____   _____

EUREKA MATH™

Lesson 10:   Estimate and measure liquid volume in liters and milliliters using the vertical number line.

©2018 Great Minds®. eureka-math.org

127

4. Kristen is comparing the capacity of gas tanks in different size cars. Use the chart below to answer the questions.

Size of Car	Capacity in Liters
Large	74
Medium	57
Small	42

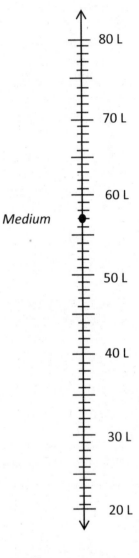

a. Label the number line to show the capacity of each gas tank. The medium car has been done for you.

b. Which car's gas tank has the greatest capacity?

c. Which car's gas tank has the smallest capacity?

d. Kristen's car has a gas tank capacity of about 60 liters. Which car from the chart has about the same capacity as Kristen's car?

e. Use the number line to find how many more liters the large car's tank holds than the small car's tank.

Lesson 10: Estimate and measure liquid volume in liters and milliliters using the vertical number line.

©2018 Great Minds®. eureka-math.org

EUREKA MATH™

1. Together the weight of a banana and an apple is 291 grams. The banana weighs 136 grams. How much does the apple weigh?

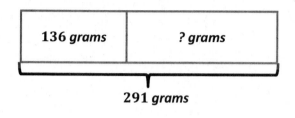

136 *grams*	*? grams*

291 *grams*

I can draw a tape diagram to model the problem. The total is 291 grams, and one part—the weight of the banana—is 136 grams. I can subtract to find the other part, the weight of the apple.

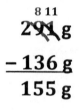

$$\begin{array}{r} \overset{8\ 11}{2\cancel{9}\cancel{1}}\text{ g} \\ -\ 136\text{ g} \\ \hline 155\text{ g} \end{array}$$

I can use the standard algorithm to subtract. I can unbundle 1 ten to make 10 ones. Now there are 2 hundreds, 8 tens, and 11 ones.

The apple weighs 155 grams.

2. Sandy uses a total of 21 liters of water to water her flowerbeds. She uses 3 liters of water for each flowerbed. How many flowerbeds does Sandy water?

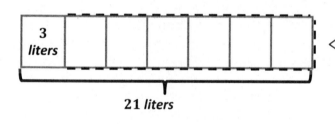

3 liters						

21 *liters*

I can draw a tape diagram to model the problem. The total is 21 liters, and each unit represents the amount of water Sandy uses for each flowerbed, 3 liters. I can see that the unknown is the number of units (groups).

$21 \div 3 = 7$

I can divide to find the total number of units, which represents the number of flowerbeds.

Sandy waters 7 flowerbeds.

Now that I know the answer, I can draw the rest of the units in my tape diagram, to show a total of 7 units.

EUREKA MATH™

Lesson 11: Solve mixed word problems involving all four operations with grams, kilograms, liters, and milliliters given in the same units.

129

©2018 Great Minds®. eureka-math.org

Name _____   Date _____

1. Karina goes on a hike. She brings a notebook, a pencil, and a camera. The weight of each item is shown in the chart. What is the total weight of all three items?

Item	Weight
Notebook	312 g
Pencil	10 g
Camera	365 g

The total weight is _____ grams.

---

2. Together a horse and its rider weigh 729 kilograms. The horse weighs 625 kilograms. How much does the rider weigh?

The rider weighs _____ kilograms.

EUREKA MATH™   **Lesson 11:**   Solve mixed word problems involving all four operations with grams, kilograms, liters, and milliliters given in the same units.   **131**

©2018 Great Minds®. eureka-math.org

3. Theresa's soccer team fills up 6 water coolers before the game. Each water cooler holds 9 liters of water. How many liters of water do they fill?

4. Dwight purchased 48 kilograms of fertilizer for his vegetable garden. He needs 6 kilograms of fertilizer for each bed of vegetables. How many beds of vegetables can he fertilize?

5. Nancy bakes 7 cakes for the school bake sale. Each cake requires 5 milliliters of oil. How many milliliters of oil does she use?

Lesson 11: Solve mixed word problems involving all four operations with grams, kilograms, liters, and milliliters given in the same units.

©2018 Great Minds®. eureka-math.org

EUREKA MATH™

1. Complete the chart.

I measured the width of a picture frame. It was 24 centimeters wide.

Object	Measurement (in cm)	The object measures between (which two tens)...	Length rounded to the nearest 10 cm
Width of picture frame	**24 cm**	___**20**___ and___**30**___cm	**20 cm**

I can use a vertical number line to help me round 24 cm to the nearest 10 cm.

The endpoints on my vertical number line help me know which two tens the width of the picture frame is in between.

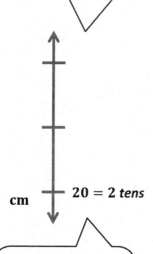

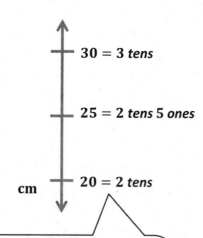

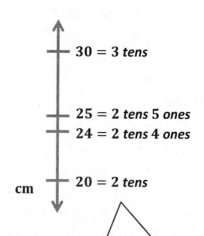

There are 2 tens in 24, so I can label this endpoint as 2 tens or 20.

One more ten than 2 tens is 3 tens, so I can label the other endpoint as 3 tens or 30. Halfway between 2 tens and 3 tens is 2 tens 5 ones. I can label the halfway point as 2 tens 5 ones or 25.

I can plot 24 or 2 tens 4 ones on the vertical number line. I can easily see that 24 is less than halfway between 2 tens and 3 tens. That means that 24 cm rounded to the nearest 10 cm is 20 cm.

EUREKA MATH™

Lesson 12: Round two-digit measurements to the nearest ten on the vertical number line.

©2018 Great Minds®. eureka-math.org

133

2. Measure the liquid in the beaker to the nearest 10 milliliters.

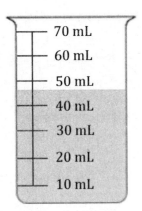

I can use the beaker to help me round the amount of liquid to the nearest 10 mL. I can see that the liquid is between 40 (4 tens) and 50 (5 tens). I can also see that the liquid is more than halfway between 4 tens and 5 tens. That means that the amount of liquid rounds up to the next ten milliliters, 50 mL.

There are about___**50**___ milliliters of liquid in the beaker.

The word *about* tells me that this is not the exact amount of liquid in the beaker.

Lesson 12:   Round two-digit measurements to the nearest ten on the vertical number line.

EUREKA MATH™

Name _____ Date _____

1. Complete the chart. Choose objects, and use a ruler or meter stick to complete the last two on your own.

Object	Measurement (in cm)	The object measures between (which two tens)...	Length rounded to the nearest 10 cm
Length of desk	66 cm	_____ and _____ cm	
Width of desk	48 cm	_____ and _____ cm	
Width of door	81 cm	_____ and _____ cm	
		_____ and _____ cm	
		_____ and _____ cm	

2. Gym class ends at 10:27 a.m. Round the time to the nearest 10 minutes.

Gym class ends at about _____ a.m.

3. Measure the liquid in the beaker to the nearest 10 milliliters.

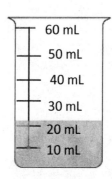

There are about _____ milliliters in the beaker.

EUREKA MATH

Lesson 12: Round two-digit measurements to the nearest ten on the vertical number line.

135

©2018 Great Minds®. eureka-math.org

4. Mrs. Santos' weight is shown on the scale. Round her weight to the nearest 10 kilograms.

Mrs. Santos' weight is _____ kilograms.

Mrs. Santos weighs about _____ kilograms.

5. A zookeeper weighs a chimp. Round the chimp's weight to the nearest 10 kilograms.

The chimp's weight is _____ kilograms.

The chimp weighs about _____ kilograms.

Lesson 12:   Round two-digit measurements to the nearest ten on the vertical number line.

EUREKA MATH

1. Round to the nearest ten. Draw a number line to model your thinking.

   a. $52 \approx$ __50__

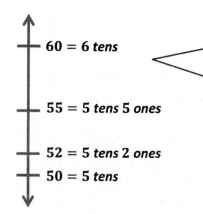

   60 = 6 *tens*

   55 = 5 *tens* 5 *ones*

   52 = 5 *tens* 2 *ones*
   50 = 5 *tens*

   > I can draw a vertical number line with endpoints of 50 and 60 and a halfway point of 55. When I plot 52 on the vertical number line, I can see that it is less than halfway between 50 and 60. So 52 rounded to the nearest ten is 50.

   b. $152 \approx$ __150__

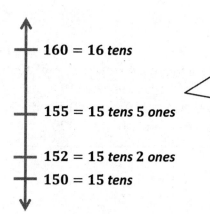

   160 = 16 *tens*

   155 = 15 *tens* 5 *ones*

   152 = 15 *tens* 2 *ones*
   150 = 15 *tens*

   > I can draw a vertical number line with endpoints of 150 and 160 and a halfway point of 155. When I plot 152 on the vertical number line, I can see that it is less than halfway between 150 and 160. So 152 rounded to the nearest ten is 150.

   > Look, my vertical number lines for parts (a) and (b) are almost the same! The only difference is that all the numbers in part (b) are 100 more than the numbers in part (a).

EUREKA MATH™

Lesson 13: Round two- and three-digit numbers to the nearest ten on the vertical number line.

©2018 Great Minds®. eureka-math.org

137

2. Amelia pours 63 mL of water into a beaker. Madison pours 56 mL of water into Amelia's beaker. Round the total amount of water in the beaker to the nearest 10 milliliters. Model your thinking using a number line.

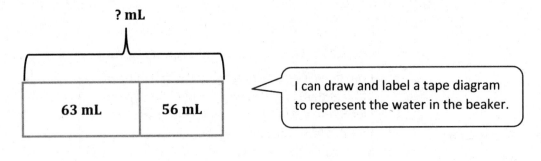

? mL

63 mL	56 mL

I can draw and label a tape diagram to represent the water in the beaker.

63 mL + 56 mL = 119 mL

I can find the total amount of water in the beaker by adding 63 mL and 56 mL.

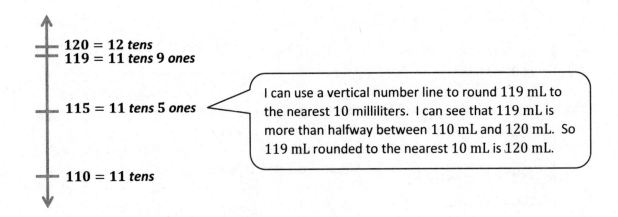

120 = 12 *tens*
119 = 11 *tens* 9 *ones*

115 = 11 *tens* 5 *ones*

110 = 11 *tens*

I can use a vertical number line to round 119 mL to the nearest 10 milliliters. I can see that 119 mL is more than halfway between 110 mL and 120 mL. So 119 mL rounded to the nearest 10 mL is 120 mL.

*There are about **120 mL** of water in the beaker.*

Lesson 13: Round two- and three-digit numbers to the nearest ten on the vertical number line.

©2018 Great Minds®. eureka-math.org

EUREKA
MATH

Name _____   Date _____

1. Round to the nearest ten. Use the number line to model your thinking.

a.   43 ≈ _____

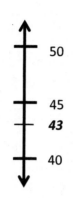

50

45

43

40

b.   48 ≈ _____

c.   73 ≈ _____

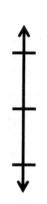

d.   173 ≈ _____

e.   189 ≈ _____

f.   194 ≈ _____

EUREKA MATH™

Lesson 13:   Round two- and three-digit numbers to the nearest ten on the vertical number line.

©2018 Great Minds®. eureka-math.org

139

2. Round the weight of each item to the nearest 10 grams. Draw number lines to model your thinking.

Item	Number Line	Round to the nearest 10 grams
Cereal bar: 45 grams		
Loaf of bread: 673 grams		

3. The Garden Club plants rows of carrots in the garden. One seed packet weighs 28 grams. Round the total weight of 2 seed packets to the nearest 10 grams. Model your thinking using a number line.

Lesson 13:   Round two- and three-digit numbers to the nearest ten on the vertical number line.

©2018 Great Minds®. eureka-math.org

EUREKA MATH™

1. Round to the nearest hundred. Draw a number line to model your thinking.

a. 234 ≈ __**200**__

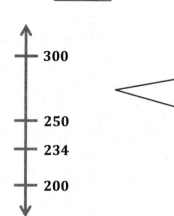

I can draw a vertical number line with endpoints of 200 and 300 and a halfway point of 250. When I plot 234 on the vertical number line, I can see that it is less than halfway between 200 and 300. So 234 rounded to the nearest hundred is 200.

b. 1,234 ≈ __**1,200**__

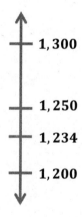

I can draw a vertical number line with endpoints of 1,200 and 1,300 and a halfway point of 1,250. When I plot 1,234 on the vertical number line, I can see that it is less than halfway between 1,200 and 1,300. So 1,234 rounded to the nearest hundred is 1,200.

Look, my vertical number lines for parts (a) and (b) are almost the same! The only difference is that all the numbers in part (b) are 1,000 more than the numbers in part (a).

EUREKA
MATH™

Lesson 14: Round to the nearest hundred on the vertical number line.

141

©2018 Great Minds®. eureka-math.org

2. There are 1,365 students at Park Street School. Kate and Sam round the number of students to the nearest hundred. Kate says it is one thousand, four hundred. Sam says it is 14 hundreds. Who is correct? Explain your thinking.

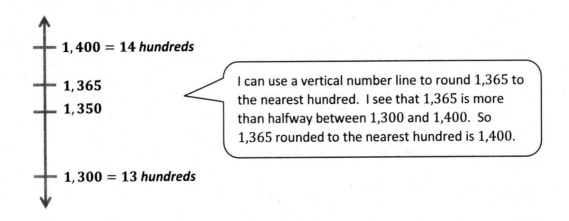

I can use a vertical number line to round 1,365 to the nearest hundred. I see that 1,365 is more than halfway between 1,300 and 1,400. So 1,365 rounded to the nearest hundred is 1,400.

Kate and Sam are both right. 1,365 rounded to the nearest hundred is 1,400. 1,400 in unit form is 14 hundreds.

        Lesson 14:   Round to the nearest hundred on the vertical number line.

EUREKA
MATH™

Name _____   Date _____

1. Round to the nearest hundred. Use the number line to model your thinking.

a. 156 ≈ _____ 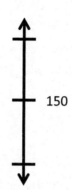 150	b. 342 ≈ _____ 
c. 260 ≈ _____ 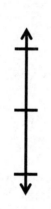	d. 1,260 ≈ _____ 
e. 1,685 ≈ _____ 	f. 1,804 ≈ _____ 

EUREKA
MATH™

Lesson 14:   Round to the nearest hundred on the vertical number line.

143

©2018 Great Minds®. eureka-math.org

2. Complete the chart.

a. Luis has 217 baseball cards. Round the number of cards Luis has to the nearest hundred.	
b. There were 462 people sitting in the audience. Round the number of people to the nearest hundred.	
c. A bottle of juice holds 386 milliliters. Round the capacity to the nearest 100 milliliters.	
d. A book weighs 727 grams. Round the weight to the nearest 100 grams.	
e. Joanie's parents spent $1,260 on two plane tickets. Round the total to the nearest $100.	

3. Circle the numbers that round to 400 when rounding to the nearest hundred.

368	342	420	492	449	464

4. There are 1,525 pages in a book. Julia and Kim round the number of pages to the nearest hundred. Julia says it is one thousand, five hundred. Kim says it is 15 hundreds. Who is correct? Explain your thinking.

Lesson 14: Round to the nearest hundred on the vertical number line.

EUREKA MATH

1. Find the sums below. Choose mental math or the algorithm.

a. 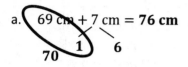 69 cm + 7 cm = **76 cm**

      70   1   6

> I can use mental math to solve this problem. I broke apart the 7 as 1 and 6. Then I solved the equation as 70 cm + 6 cm = 76 cm.

> For this problem, the standard algorithm is a more strategic tool to use.

b.   59 kg + 76 kg

$$\begin{array}{r} 59 \text{ kg} \\ +\ 76 \text{ kg} \\ \hline {\scriptstyle 1} \\ 5 \end{array}$$

$$\begin{array}{r} 59 \text{ kg} \\ +\ 76 \text{ kg} \\ \hline {\scriptstyle 1} \\ 135 \text{ kg} \end{array}$$

> 9 ones plus 6 ones is 15 ones. I can rename 15 ones as 1 ten and 5 ones. I can record this by writing the 1 so that it crosses the line under the tens in the tens place, and the 5 below the line in the ones column. This way I write 15, rather than 5 and 1 as separate numbers.

> 5 tens plus 7 tens plus 1 ten equals 13 tens. So, 59 kg + 76 kg = 135 kg.

EUREKA MATH™

Lesson 15: Add measurements using the standard algorithm to compose larger units once.

©2018 Great Minds®. eureka-math.org

145

2. Mrs. Alvarez's plant grew 23 centimeters in one week. The next week it grew 6 centimeters more than the previous week. What is the total number of centimeters the plant grew in 2 weeks?

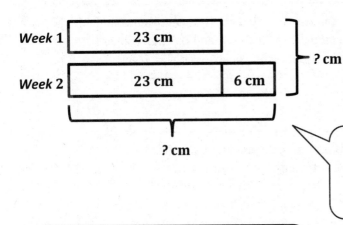

I can draw a double tape diagram for this problem because I am comparing Week 1 and Week 2.

I know that in Week 2 the plant grew 6 centimeters more than the previous week. So, I can add on 6 cm to 23 cm to get 29 cm in Week 2.

29 cm does not answer the question since this tells me how much the plant grew only in Week 2. I need to find the total number of centimeters the plant grew in 2 weeks.

$$23 \text{ cm} + 6 \text{ cm} = 29 \text{ cm}$$

In order to find the total number of centimeters the plant grew in 2 weeks, I can add 23 cm + 29 cm. I can use mental math to solve this problem since 29 is close to 30.

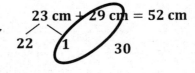

$$23 \text{ cm} + 29 \text{ cm} = 52 \text{ cm}$$
22   1   30

Now I can write a statement that answers the question. This helps me check my work to see if my answer is reasonable.

The plant grew 52 centimeters in 2 weeks.

Lesson 15:   Add measurements using the standard algorithm to compose larger units once.

©2018 Great Minds®. eureka-math.org

EUREKA
MATH™

Name _____ Date _____

1. Find the sums below. Choose mental math or the algorithm.

   a. 75 cm + 7 cm

   c. 362 mL + 229 mL

   e. 451 mL + 339 mL

   b. 39 kg + 56 kg

   d. 283 g + 92 g

   f. 149 L + 331 L

2. The liquid volume of five drinks is shown below.

Drink	Liquid Volume
Apple juice	125 mL
Milk	236 mL
Water	248 mL
Orange juice	174 mL
Fruit punch	208 mL

   a. Jen drinks the apple juice and the water. How many milliliters does she drink in all?

   Jen drinks _____ mL.

   b. Kevin drinks the milk and the fruit punch. How many milliliters does he drink in all?

EUREKA MATH™

Lesson 15: Add measurements using the standard algorithm to compose larger units once.

©2018 Great Minds®. eureka-math.org

147

3. There are 75 students in Grade 3. There are 44 more students in Grade 4 than in Grade 3. How many students are in Grade 4?

4. Mr. Green's sunflower grew 29 centimeters in one week. The next week it grew 5 centimeters more than the previous week. What is the total number of centimeters the sunflower grew in 2 weeks?

5. Kylie records the weights of 3 objects as shown below. Which 2 objects can she put on a pan balance to equal the weight of a 460 gram bag? Show how you know.

Paperback Book	Banana	Bar of Soap
343 grams	108 grams	117 grams

Lesson 15:   Add measurements using the standard algorithm to compose larger
             units once.

EUREKA
MATH™

1. Find the sums.

a. 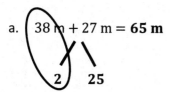 $38 \text{ m} + 27 \text{ m} = \mathbf{65 \text{ m}}$

> I can use mental math to solve this problem. I can break apart 27 as 2 and 25. Then I can solve 40 m + 25 m, which is 65 m.

b. 358 kg + 167 kg

> I can use the standard algorithm to solve this problem. I can line the numbers up vertically and add.

385 kg	**385 kg**	**385 kg**
+ 167 kg	**+ 167 kg**	**+ 167 kg**
$\overset{1}{}$	$\overset{11}{}$	$\overset{11}{}$
2	**52**	**552 kg**

> 5 ones plus 7 ones is 12 ones. I can rename 12 ones as 1 ten 2 ones.

> 8 tens plus 6 tens is 14 tens. Plus 1 more ten is 15 tens. I can rename 15 tens as 1 hundred 5 tens.

> 3 hundreds plus 1 hundred is 4 hundreds. Plus 1 more hundred is 5 hundreds. The sum is 552 kg.

EUREKA MATH™

Lesson 16: Add measurements using the standard algorithm to compose larger units twice.

©2018 Great Minds®. eureka-math.org

149

2. Matthew reads for 58 more minutes in March than in April. He reads for 378 minutes in April. Use a tape diagram to find the total minutes Matthew reads in March and April.

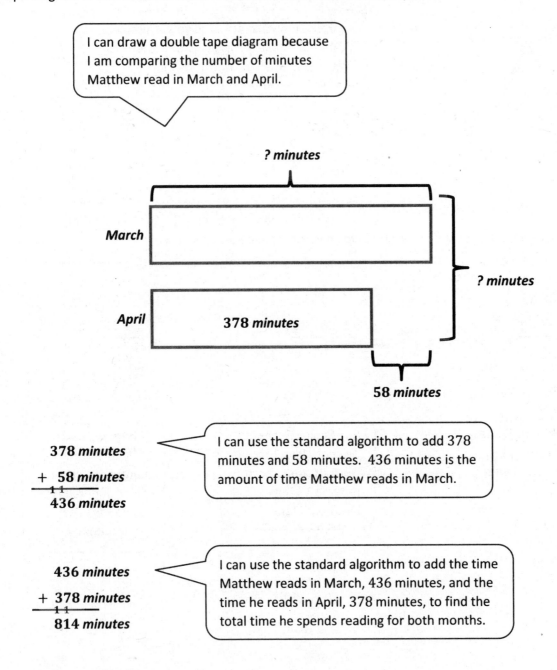

I can draw a double tape diagram because I am comparing the number of minutes Matthew read in March and April.

? minutes

March

April   378 minutes

? minutes

58 minutes

378 minutes
+ 58 minutes
   1 1
436 minutes

I can use the standard algorithm to add 378 minutes and 58 minutes. 436 minutes is the amount of time Matthew reads in March.

436 minutes
+ 378 minutes
  1 1
814 minutes

I can use the standard algorithm to add the time Matthew reads in March, 436 minutes, and the time he reads in April, 378 minutes, to find the total time he spends reading for both months.

Matthew read for 814 minutes in March and April.

Lesson 16:   Add measurements using the standard algorithm to compose larger units twice.

EUREKA MATH

Lucy buys an apple that weighs 152 grams. She buys a banana that weighs 109 gram

a. Estimate the total weight of the apple and banana by rounding.

$152 \approx 200$

$109 \approx 100$

> I can round each number to the nearest hundred.

$200\ grams + 100\ grams = 300\ grams$

> I can add the rounded numbers to e   nate the total weight of the apple and the nana. The total weight is about 300 grams.

b. Estimate the total weight of the apple and banana by rounding in a different way.

$152 \approx 150$

$109 \approx 110$

> I can round each number to the nearest ten.

$150\ grams + 110\ grams = 260\ grams$

> I can add the rounded numbers to estimate the total weight of the apple and the banana. The total weight is about 260 grams.

c. Calculate the actual total weight of the apple and the banana. Which method of rounding was more precise? Why?

$152\ grams$

$+\ 109\ grams$

$261\ grams$

Rounding to the nearest ten grams was more precise because when I rounded to the nearest ten grams, the estimate was 260 grams, and the actual answer is 261 grams. The estimate and the actual answer are only 1 gram apart! When I rounded to the nearest hundred grams, the estimate was 300 grams, which isn't that close to the actual answer.

> I can use the standard algorithm to find the actual total weight of the apple and the banana.

Name _____   Date _____

1. Cathy collects the following information about her dogs, Stella and Oliver.

Stella	
Time Spent Getting a Bath	Weight
36 minutes	32 kg

Oliver	
Time Spent Getting a Bath	Weight
25 minutes	7 kg

Use the information in the charts to answer the questions below.

a. Estimate the total weight of Stella and Oliver.

b. What is the actual total weight of Stella and Oliver?

c. Estimate the total amount of time Cathy spends giving her dogs a bath.

d. What is the actual total time Cathy spends giving her dogs a bath?

e. Explain how estimating helps you check the reasonableness of your answers.

EUREKA
MATH™

Lesson 17:   Estimate sums by rounding and apply to solve measurement word problems.

©2018 Great Minds®. eureka-math.org

155

2. Dena reads for 361 minutes during Week 1 of her school's two-week long Read-A-Thon. She reads for 212 minutes during Week 2 of the Read-A-Thon.

   a. Estimate the total amount of time Dena reads during the Read-A-Thon by rounding.

   b. Estimate the total amount of time Dena reads during the Read-A-Thon by rounding in a different way.

   c. Calculate the actual number of minutes that Dena reads during the Read-A-Thon. Which method of rounding was more precise? Why?

Lesson 17:   Estimate sums by rounding and apply to solve measurement word
                       problems.

       ©2018 Great Minds®. eureka-math.org

EUREKA
MATH™

1. Solve the subtraction problems below.

a. $50 \text{ cm} - 24 \text{ cm} = \textbf{26 cm}$

> I can use mental math to solve this subtraction problem. I do not have to write it out vertically. I can also think of my work with quarters. I know $50 - 25 = 25$. But since I'm only subtracting 24, I need to add 1 more to 25. So, the answer is 26 cm.

b. $507 \text{ g} - 234 \text{ g}$

$$\begin{array}{r} 507 \text{ g} \\ - 234 \text{ g} \\ \hline \end{array}$$

> Before I subtract, I need to see if any tens or hundreds need to be unbundled. I can see that there are enough ones to subtract 4 ones from 7 ones. There is no need to unbundle a ten.

$$\begin{array}{r} \overset{4\ 10}{\cancel{5}\cancel{0}7} \text{ g} \\ - 234 \text{ g} \\ \hline \end{array}$$

> But, I am still not ready to subtract. There are not enough tens to subtract 3 tens, so I need to unbundle 1 hundred to make 10 tens. Since I unbundled 1 hundred, there are now 4 hundreds left.

$$\begin{array}{r} \overset{4\ 10}{\cancel{5}\cancel{0}7} \text{ g} \\ - 234 \text{ g} \\ \hline 273 \text{ g} \end{array}$$

> After unbundling, I see that there are 4 hundreds, 10 tens, and 7 ones. Now I am ready to subtract. Since I've prepared my numbers all at once, I can subtract left to right, or right to left. The answer is 273 grams.

EUREKA MATH

Lesson 18: Decompose once to subtract measurements including three-digit minuends with zeros in the tens or ones place.

©2018 Great Minds®. eureka-math.org

157

2. Renee buys 607 grams of cherries at the market on Monday. On Wednesday, she buys 345 grams of cherries. How many more grams of cherries did Renee buy on Monday than on Wednesday?

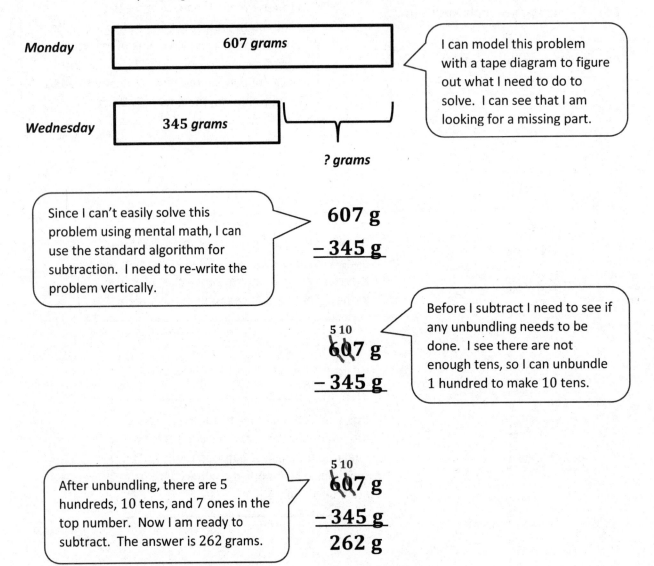

Monday
607 grams

I can model this problem with a tape diagram to figure out what I need to do to solve. I can see that I am looking for a missing part.

Wednesday
345 grams

? grams

Since I can't easily solve this problem using mental math, I can use the standard algorithm for subtraction. I need to re-write the problem vertically.

607 g
− 345 g

Before I subtract I need to see if any unbundling needs to be done. I see there are not enough tens, so I can unbundle 1 hundred to make 10 tens.

5 10
6̶0̶7 g
− 345 g

After unbundling, there are 5 hundreds, 10 tens, and 7 ones in the top number. Now I am ready to subtract. The answer is 262 grams.

5 10
6̶0̶7 g
− 345 g
262 g

Renee buys 262 more grams of cherries on Monday than on Wednesday.

Lesson 18:   Decompose once to subtract measurements including three-digit minuends with zeros in the tens or ones place.

EUREKA MATH™

Name _____   Date _____

1. Solve the subtraction problems below.

   a. 70 L – 46 L

   b. 370 L – 46 L

   c. 370 L – 146 L

   d. 607 cm – 32 cm

   e. 592 cm – 258 cm

   f. 918 cm – 553 cm

   g. 763 g – 82 g

   h. 803 g – 542 g

   i. 572 km – 266 km

   j. 837 km – 645 km

EUREKA
MATH™

Lesson 18:   Decompose once to subtract measurements including three-digit
             minuends with zeros in the tens or ones place.

©2018 Great Minds®. eureka-math.org

159

2. The magazine weighs 280 grams less than the newspaper. The weight of the newspaper is shown below. How much does the magazine weigh? Use a tape diagram to model your thinking.

 **454 g**

3. The chart to the right shows how long it takes to play 3 games.

   a. Francesca's basketball game is 22 minutes shorter than Lucas's baseball game. How long is Francesca's basketball game?

Lucas's Baseball Game	180 minutes
Joey's Football Game	139 minutes
Francesca's Basketball Game	? minutes

   b. How much longer is Francesca's basketball game than Joey's football game?

Lesson 18:   Decompose once to subtract measurements including three-digit minuends with zeros in the tens or ones place.

EUREKA MATH™

1. Solve the subtraction problems below.

a. 370 cm − 90 cm = **280 cm**

> I can use mental math to solve this subtraction problem. I do not have to write it out vertically. Using the compensation strategy, I can add 10 to both numbers and think of the problem as 380 − 100, which is an easy calculation. The answer is 280 cm.

b. 800 mL − 126 mL

$$
\begin{array}{r}
\overset{7\ 10}{\cancel{800}} \text{ mL} \\
- 126 \text{ mL}
\end{array}
$$

> Before I subtract, I need to see if any tens or hundreds need to be unbundled. There are not enough ones to subtract, so I can unbundle 1 ten to make 10 ones. But there are 0 tens, so I can unbundle 1 hundred to make 10 tens. Then there are 7 hundreds and 10 tens.

$$
\begin{array}{r}
\overset{\quad 9}{\overset{7\ 10\,10}{\cancel{800}}} \text{ mL} \\
- 126 \text{ mL}
\end{array}
$$

> I still am not ready to subtract because I have to unbundle 1 ten to make 10 ones. Then there are 9 tens and 10 ones.

$$
\begin{array}{r}
\overset{\quad 9}{\overset{7\ 10\,10}{\cancel{800}}} \text{ mL} \\
- 126 \text{ mL} \\
\hline
674 \text{ mL}
\end{array}
$$

> After unbundling, I see that I have 7 hundreds, 9 tens, and 10 ones. Now I am ready to subtract. Since I've prepared my numbers all at once, I can choose to subtract left to right, or right to left. The answer is 674 mL.

EUREKA MATH

Lesson 19: Decompose twice to subtract measurements including three-digit minuends with zeros in the tens and ones places.

©2018 Great Minds®. eureka-math.org

161

2. Kenny is driving from Los Angeles to San Diego. The total distance is about 175 kilometers. He has 86 kilometers left to drive. How many kilometers has he driven so far?

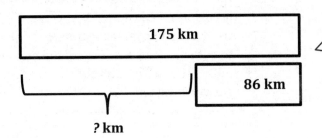

I can model this problem with a tape diagram to figure out what I need to do to solve. I can see that I am looking for a missing part.

Since I can't easily solve this problem using mental math, I can use the standard algorithm for subtraction. I can re-write the problem vertically.

175 km
− 86 km

0 17
1̶7̶5 km
− 86 km

Before I subtract, I need to see if any unbundling needs to be done. I can see there are not enough tens or ones, so I can unbundle 1 hundred to make 10 tens. After unbundling, there are 0 hundreds and 17 tens.

16
0 1̶7̶ 15
1̶7̶5̶ km
− 86 km
89 km

I can unbundle 1 ten to make 10 ones. After unbundling, there are 0 hundreds, 16 tens, and 15 ones. I am ready to subtract. The answer is 89 kilometers.

Kenny has driven 89 km so far.

Lesson 19: Decompose twice to subtract measurements including three-digit minuends with zeros in the tens and ones places.

EUREKA MATH™

Name _____   Date _____

1. Solve the subtraction problems below.

   a. 280 g – 90 g

   b. 450 g – 284 g

   c. 423 cm – 136 cm

   d. 567 cm – 246 cm

   e. 900 g – 58 g

   f. 900 g – 358 g

   g. 4 L 710 mL – 2 L 690 mL

   h. 8 L 830 mL – 4 L 378 mL

EUREKA
MATH™

Lesson 19:   Decompose twice to subtract measurements including three-digit
minuends with zeros in the tens and ones places.

©2018 Great Minds®. eureka-math.org

163

2. The total weight of a giraffe and her calf is 904 kilograms. How much does the calf weigh? Use a tape diagram to model your thinking.

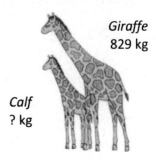

Giraffe
829 kg

Calf
? kg

3. The Erie Canal runs 584 kilometers from Albany to Buffalo. Salvador travels on the canal from Albany. He must travel 396 kilometers more before he reaches Buffalo. How many kilometers has he traveled so far?

4. Mr. Nguyen fills two inflatable pools. The kiddie pool holds 185 liters of water. The larger pool holds 600 liters of water. How much more water does the larger pool hold than the kiddie pool?

Lesson 19: Decompose twice to subtract measurements including three-digit minuends with zeros in the tens and ones places.

©2018 Great Minds®. eureka-math.org

EUREKA
MATH

Esther measures rope. She measures a total of 548 centimeters of rope and cuts it into two pieces. The first piece is 152 centimeters long. How long is the second piece of rope?

a. Estimate the length of the second piece of rope by rounding.

548 cm ≈ 500 cm

152 cm ≈ 200 cm

> I can round each number to the nearest hundred for my first estimate. I notice that both numbers are far from the hundred.

500 cm − 200 cm = 300 cm

The second piece of rope is about **300 cm** *long.*

b. Estimate the length of the second piece of ribbon by rounding in a different way.

548 cm ≈ 550 cm

152 cm ≈ 150 cm

> I can round each number to the nearest ten for my second estimate. Wow, both numbers are close to the fifty! This makes it easy to calculate.

550 cm − 150 cm = 400 cm

The second piece of rope is about **400 cm** *long.*

c. Precisely how long is the second piece of rope?

$$\begin{array}{r} \overset{4\ \ 14}{\cancel{5}\cancel{4}8}\text{ cm} \\ -\ 152\text{ cm} \\ \hline 396\text{ cm} \end{array}$$

> Before I am ready to subtract, I can unbundle 1 hundred for 10 tens.

The second piece of rope is precisely **396 cm** *long.*

d.   Is your answer reasonable? Which estimate was closer to the exact answer?

Rounding to the nearest ten was closer to the exact answer, and it was easy mental math. The estimate was only 4 cm away from the actual answer. So that's how I know my answer is reasonable.

Comparing my actual answer with my estimate helps me check my calculation because if the answers are very different, I've probably made a mistake in my calculation.

Lesson 20:   Estimate differences by rounding and apply to solve measurement
                                    word problems.

                    ©2018 Great Minds®. eureka-math.org

EUREKA
MATH™

Name _____   Date _____

Estimate, and then solve each problem.

1. Melissa and her mom go on a road trip. They drive 87 kilometers before lunch. They drive 59 kilometers after lunch.

   a. Estimate how many more kilometers they drive before lunch than after lunch by rounding to the nearest 10 kilometers.

   b. Precisely how much farther do they drive before lunch than after lunch?

   c. Compare your estimate from (a) to your answer from (b). Is your answer reasonable? Write a sentence to explain your thinking.

2. Amy measures ribbon. She measures a total of 393 centimeters of ribbon and cuts it into two pieces. The first piece is 184 centimeters long. How long is the second piece of ribbon?

   a. Estimate the length of the second piece of ribbon by rounding in two different ways.

   b. Precisely how long is the second piece of ribbon? Explain why one estimate was closer.

Lesson 20:   Estimate differences by rounding and apply to solve measurement word problems.

©2018 Great Minds®. eureka-math.org

167

3. The weight of a chicken leg, steak, and ham are shown to the
    right. The chicken and the steak together weigh 341 grams.
    How much does the ham weigh?

    a. Estimate the weight of the ham by rounding.

    989 grams

    b. How much does the ham actually weigh?

4. Kate uses 506 liters of water each week to water plants. She uses 252 liters to water the plants in the
    greenhouse. How much water does she use for the other plants?

    a. Estimate how much water Kate uses for the other plants by rounding.

    b. Estimate how much water Kate uses for the other plants by rounding a different way.

    c. How much water does Kate actually use for the other plants? Which estimate was closer? Explain
        why.

Lesson 20:   Estimate differences by rounding and apply to solve measurement
              word problems.

©2018 Great Minds®. eureka-math.org

EUREKA
MATH™

Mia measures the lengths of three pieces of wire. The lengths of the wires are recorded to the right.

Wire A	63 cm ≈ __60__ cm
Wire B	75 cm ≈ __80__ cm
Wire C	49 cm ≈ __50__ cm

a. Estimate the total length of Wire A and Wire C. Then, find the actual total length.

> I can round the lengths of all the wires to the nearest ten.

Estimate: **60 cm + 50 cm = 110 cm**

> I can add the rounded lengths of Wires A and C to find an estimate of their total length.

Actual: **63 cm + 49 cm = 112 cm**

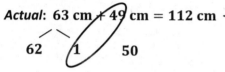

> I can use mental math to solve this problem. I do not have to write it out vertically. I can break apart 63 as 62 and 1. Then I can make the next ten to 50, and then add the 62.

The total length is 112 cm.

b. Subtract to estimate the difference between the total length of Wires A and C and the length of Wire B. Then, find the actual difference. Model the problem with a tape diagram.

Estimate: **110 cm − 80 cm = 30 cm**

Actual: **112 cm − 75 cm = 37 cm**

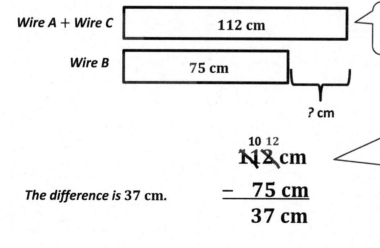

> From the tape diagram, I see that I need to solve for an unknown part.

The difference is 37 cm.

$$
\begin{array}{r}
\overset{10}{\cancel{11}}\overset{12}{\cancel{2}}\text{ cm} \\
-\quad 75\text{ cm} \\
\hline
37\text{ cm}
\end{array}
$$

> I can write this problem vertically. I can unbundle 1 ten for 10 ones. I can rename 112 as 10 tens and 12 ones. Then I am ready to subtract.

EUREKA MATH

Lesson 21: Estimate sums and differences of measurements by rounding, and then solve mixed word problems.

169

©2018 Great Minds®. eureka-math.org

Name _____ Date _____

1. There are 153 milliliters of juice in 1 carton. A three-pack of juice boxes contains a total of 459 milliliters.

   a. Estimate, and then find the actual total amount of juice in 1 carton and in a three-pack of juice boxes.

      153 mL + 459 mL ≈ _____ + _____ = _____

      153 mL + 459 mL = _____

   b. Estimate, and then find the actual difference between the amount in 1 carton and in a three-pack of juice boxes.

      459 mL – 153 mL ≈ _____ – _____ = _____

      459 mL – 153 mL = _____

   c. Are your answers reasonable? Why?

2. Mr. Williams owns a gas station. He sells 367 liters of gas in the morning, 300 liters of gas in the afternoon, and 219 liters of gas in the evening.

   a. Estimate, and then find the actual total amount of gas he sells in one day.

   b. Estimate, and then find the actual difference between the amount of gas Mr. Williams sells in the morning and the amount he sells in the evening.

EUREKA MATH™   Lesson 21:   Estimate sums and differences of measurements by rounding, and then solve mixed word problems.   171

©2018 Great Minds®. eureka-math.org

3. The Blue Team runs a relay. The chart shows the time, in minutes, that each team member spends running.

Blue Team	Time in Minutes
Jen	5 minutes
Kristin	7 minutes
Lester	6 minutes
Evy	8 minutes
Total	

a. How many minutes does it take the Blue Team to run the relay?

b. It takes the Red Team 37 minutes to run the relay. Estimate, and then find the actual difference in time between the two teams.

4. The lengths of three banners are shown to the right.

Banner A	437 cm
Banner B	457 cm
Banner C	332 cm

a. Estimate, and then find the actual total length of Banner A and Banner C.

b. Estimate, and then find the actual difference in length between Banner B and the combined length of Banner A and Banner C. Model the problem with a tape diagram.

Lesson 21: Estimate sums and differences of measurements by rounding, and then solve mixed word problems.

©2018 Great Minds®. eureka-math.org

EUREKA
MATH

Grade 3
Module 3

1. Write two multiplication facts for each array.

This array shows 3 rows of 7 dots, or 3 sevens. 3 sevens can be written as $3 \times 7 = 21$. I can also write it as $7 \times 3 = 21$ using the commutative property.

__21__ = __3__ × __7__

__21__ = __7__ × __3__

2. Match the expressions.

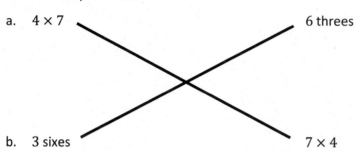

a. 4×7 6 threes

b.   3 sixes 7×4

The commutative property says that even if the order of the factors changes, the product stays the same!

3. Complete the equations.

a. $7 \times$ __2__ = __7__ $\times 2$

= __14__

This equation shows that both sides equal the same amount. Since the factors 7 and 2 are already given, I just have to fill in the unknowns with the correct factors to show that each side equals 14.

b.   6 twos + 2 twos = __8__ × __2__

= __16__

This equation shows the break apart and distribute strategy that I learned in Module 1. 6 twos + 2 twos = 8 twos, or 8×2. Since I know $2 \times 8 = 16$, I also know $8 \times 2 = 16$ using commutativity. Using commutativity as a strategy allows me to know many more facts than the ones I've practiced before.

EUREKA MATH™

Lesson 1:   Study commutativity to find known facts of 6, 7, 8, and 9.

175

©2018 Great Minds®. eureka-math.org

Name _____ Date _____

1. Complete the charts below.

    a. A tricycle has 3 wheels.

Number of Tricycles	3		5		7
Total Number of Wheels		12		18	

    b. A tiger has 4 legs.

Number of Tigers			7	8	9
Total Number of Legs	20	24			

    c. A package has 5 erasers.

Number of Packages	6				10
Total Number of Erasers		35	40	45	

2. Write two multiplication facts for each array.

_____ = _____ × _____

_____ = _____ × _____

_____ = _____ × _____

_____ = _____ × _____

EUREKA MATH™

Lesson 1: Study commutativity to find known facts of 6, 7, 8, and 9.

177

©2018 Great Minds®. eureka-math.org

3. Match the expressions.

3 × 6   7 threes

3 sevens   2 × 10

2 eights   9 × 5

5 × 9   8 × 2

10 twos   6 × 3

---

4. Complete the equations.

a.   2 sixes = _____ twos   d.   4 × _____ = _____ × 4

= __12__   = __28__

b.   _____ × 6 = 6 threes   e.   5 twos + 2 twos = _____ × _____

= _____   = _____

c.   4 × 8 = _____ × 4   f.   _____ fives + 1 five = 6 × 5

= _____   = _____

EUREKA
MATH

1. Each 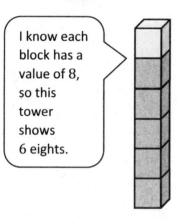 has a value of 8.

> I know each block has a value of 8, so this tower shows 6 eights.

Unit form: 6 eights = __5__ eights + __1__ eight

$$= 40 + \underline{8}$$

$$= \underline{48}$$

> The shaded and unshaded blocks show 6 eights broken into 5 eights and 1 eight. These two smaller facts will help me solve the larger fact.

Facts:

__6__ × __8__ = __48__

__8__ × __6__ = __48__

> Using commutativity, I can solve 2 multiplication facts, 6 × 8 and 8 × 6, which both equal 48.

2. There are 7 blades on each pinwheel. How many total blades are on 8 pinwheels? Use a fives fact to solve.

> I need to find the value of 8 × 7, or 8 sevens. I can draw a picture. Each dot has a value of 7. I can use my familiar fives facts to break up 8 sevens as 5 sevens and 3 sevens.

$$8 \times 7 = (5 \times 7) + (3 \times 7)$$

$$= 35 + 21$$

$$= 56$$

> This is how I write the larger fact as the sum of two smaller facts. I can add their products to find the answer to the larger fact. 8 × 7 = 56

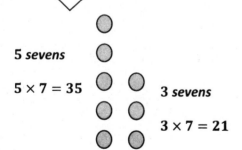

5 sevens

$5 \times 7 = 35$

3 sevens

$3 \times 7 = 21$

There are 56 blades on 8 pinwheels.

EUREKA MATH™

Lesson 2: Apply the distributive and commutative properties to relate multiplication facts $5 \times n + n$ to $6 \times n$ and $n \times 6$ where n is the size of the unit.

©2018 Great Minds®. eureka-math.org

179

Name _____   Date _____

1. Each has a value of 9.

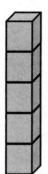

Unit form: _____

Facts: 5 × _____ = _____ × 5

Total = _____

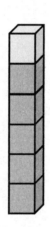

Unit form: 6 nines = _____ nines + _____ nine

= 45 + _____

= _____

Facts: _____ × _____ = _____

_____ × _____ = _____

Lesson 2: Apply the distributive and commutative properties to relate
multiplication facts 5 × n + n to 6 × n and n × 6 where n is the size
of the unit.
©2018 Great Minds®. eureka-math.org

EUREKA MATH

181

2.   There are 6 blades on each windmill. How many total blades are on 7 windmills? Use a fives fact to solve.

---

3.   Juanita organizes her magazines into 3 equal piles. She has a total of 18 magazines. How many magazines are in each pile?

---

4.   Markuo spends $27 on some plants. Each plant costs $9. How many plants does he buy?

Lesson 2:   Apply the distributive and commutative properties to relate multiplication facts $5 \times n + n$ to $6 \times n$ and $n \times 6$ where n is the size of the unit.

©2018 Great Minds®. eureka-math.org

EUREKA
MATH™

1. Each equation contains a letter representing the unknown. Find the value of the unknown.

$9 \div 3 = c$	$c = \underline{\ \ 3\ \ }$
$4 \times a = 20$	$a = \underline{\ \ 5\ \ }$

> I can think of this problem as division, $20 \div 4$, to find the unknown factor.

2. Brian buys 4 journals at the store for $8 each. What is the total amount Brian spends on 4 journals? Use the letter j to represent the total amount Brian spends, and then solve the problem.

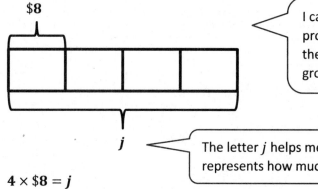

$$4 \times \$8 = j$$
$$j = \$32$$

> I can draw a tape diagram to help me solve this problem. From the diagram, I can see that I know the number of groups, 4, and the size of each group, $8, but I don't know the whole.

> The letter j helps me label the unknown, which represents how much money Brian spends on 4 journals.

Brian spends $32 on 4 journals.

> The only thing different about using a letter to solve is that I use the letter to label the unknowns in the tape diagram and in the equation. Other than that, it doesn't change the way I solve. I found the value of j is $32.

EUREKA MATH

Lesson 3:   Multiply and divide with familiar facts using a letter to represent the unknown.

©2018 Great Minds®. eureka-math.org

183

Name _____ Date _____

1. a. Complete the pattern.

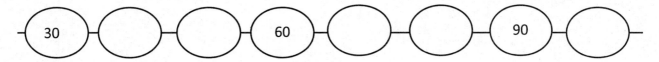

    b. Find the value of the unknown.

$10 \times 2 = d$ d = __20__ $10 \times 6 = w$ w = _____

$3 \times 10 = e$ e = _____ $10 \times 7 = n$ n = _____

$f = 4 \times 10$ f = _____ $g = 8 \times 10$ g = _____

$p = 5 \times 10$ p = _____

2. Each equation contains a letter representing the unknown. Find the value of the unknown.

$8 \div 2 = n$	n = _____
$3 \times a = 12$	a = _____
$p \times 8 = 40$	p = _____
$18 \div 6 = c$	c = _____
$d \times 4 = 24$	d = _____
$h \div 7 = 5$	h = _____
$6 \times 3 = f$	f = _____
$32 \div y = 4$	y = _____

EUREKA MATH™

Lesson 3: Multiply and divide with familiar facts using a letter to represent the unknown.

©2018 Great Minds®. eureka-math.org

185

3. Pedro buys 4 books at the fair for $7 each.

  a. What is the total amount Pedro spends on 4 books? Use the letter b to represent the total amount Pedro spends, and then solve the problem.

  b. Pedro hands the cashier 3 ten dollar bills. How much change will he receive? Write an equation to solve. Use the letter c to represent the unknown.

4. On field day, the first-grade dash is 25 meters long. The third-grade dash is twice the distance of the first-grade dash. How long is the third-grade dash? Use a letter to represent the unknown and solve.

Lesson 3:   Multiply and divide with familiar facts using a letter to represent the unknown.

©2018 Great Minds®. eureka-math.org

EUREKA
MATH

1. Use number bonds to help you skip-count by six by either making a ten or adding to the ones.

    $60 + 6 = \underline{\quad 66 \quad}$

    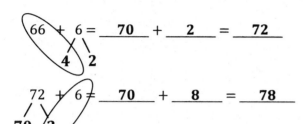

    $66 + 6 = \underline{\quad 70 \quad} + \underline{\quad 2 \quad} = \underline{\quad 72 \quad}$

    $72 + 6 = \underline{\quad 70 \quad} + \underline{\quad 8 \quad} = \underline{\quad 78 \quad}$

    > I can break apart an addend to make a ten. For example, I see that 66 just needs 4 more to make 70. So I can break 6 into 4 and 2. Then 66 + 4 = 70, plus 2 makes 72. It's much easier to add from a ten. Once I get really good at this, it'll make adding with mental math simple.

2. Count by six to fill in the blanks below.

    6, $\underline{\quad 12 \quad}$, $\underline{\quad 18 \quad}$, $\underline{\quad 24 \quad}$

    > I can skip-count to see that 4 sixes make 24.

    Complete the multiplication equation that represents your count-by.

    $6 \times \underline{\quad 4 \quad} = \underline{\quad 24 \quad}$

    > 4 sixes make 24, so 6 × 4 = 24.

    Complete the division equation that represents your count-by.

    $\underline{\quad 24 \quad} \div 6 = \underline{\quad 4 \quad}$

    > I'll use a related division fact. 6 × 4 = 24, so 24 ÷ 6 = 4.

3. Count by six to solve 36 ÷ 6. Show your work below.

    6, 12, 18, 24, 30, 36

    36 ÷ 6 = 6

    > I'll skip-count by six until I get to 36. Then I can count to find the number of sixes it takes to make 36. It takes 6 sixes, so 36 ÷ 6 = 6.

Name _____   Date _____

1. Use number bonds to help you skip-count by six by either making a ten or adding to the ones.

a. 6 + 6 = __10__ + __2__ = _____
   4   2

b. 12 + 6 = __10__ + __8__ = _____
   10   2

c. 18 + 6 = _____ + _____ = _____
    2   4

d. 24 + 6 = _____ + _____ = _____
   20   4

e. 30 + 6 = _____

f. 36 + 6 = _____ + _____ = _____
    4   2

g. 42 + 6 = _____ + _____ = _____

h. 48 + 6 = _____ + _____ = _____

i. 54 + 6 = _____ + _____ = _____

2. Count by six to fill in the blanks below.

6, _____, _____, _____, _____

Complete the multiplication equation that represents the final number in your count-by.

6 × _____ = _____

Complete the division equation that represents your count-by.

_____ ÷ 6 = _____

3. Count by six to fill in the blanks below.

6, _____, _____, _____, _____, _____

Complete the multiplication equation that represents the final number in your count-by.

6 × _____ = _____

Complete the division equation that represents your count-by.

_____ ÷ 6 = _____

4. Count by six to solve 48 ÷ 6. Show your work below.

Lesson 4:   Count by units of 6 to multiply and divide using number bonds to decompose.

EUREKA MATH

1. Use number bonds to help you skip-count by seven by either making a ten or adding to the ones.

70 + 7 = __77__

77 + 7 = __80__ + __4__ = __84__
   3   4

84 + 7 = __90__ + __1__ = __91__
   6   1

> I can break apart an addend to make a ten. For example, I see that 77 just needs 3 more to make 80. So I can break 7 into 3 and 4. Then 77 + 3 = 80, plus 4 makes 84. It's much easier to add from a ten. Once I get really good at this, it'll make adding with mental math simple.

2. Count by seven to fill in the blanks. Then use the multiplication equation to write the related division fact directly to its right.

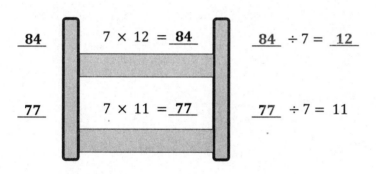

84   7 × 12 = __84__   __84__ ÷ 7 = __12__

77   7 × 11 = __77__   __77__ ÷ 7 = 11

> I "climb" the ladder counting by sevens. The count-by helps me find the products of the multiplication facts. First I find the answer to the fact on the bottom rung. I record the answer in the equation and to the left of the ladder. Then I add seven to my answer to find the next number in my count-by. The next number in my count-by is the product of the next fact up on the ladder!

> Once I find the product of a fact by skip-counting, I can write the related division fact. The total, or the product of the multiplication fact, gets divided by 7. The quotient represents the number of sevens I skip-counted.

EUREKA MATH™

Lesson 5: Count by units of 7 to multiply and divide using number bonds to decompose.

191

©2018 Great Minds®. eureka-math.org

Name _____   Date _____

1. Use number bonds to help you skip-count by seven by making ten or adding to the ones.

a.   7 + 7 = ___10___ + ___4___ = _____
         / \
       3   4

b.   14 + 7 = _____ + _____ = _____
          / \
        6   1

c.   21 + 7 = _____ + _____ = _____
         / \
      20   1

d.   28 + 7 = _____ + _____ = _____
          / \
        2   5

e.   35 + 7 = _____ + _____ = _____
          / \
        5   2

f.   42 + 7 = _____ + _____ = _____

g.   49 + 7 = _____ + _____ = _____

h.   56 + 7 = _____ + _____ = _____

2. Skip-count by seven to fill in the blanks. Then, fill in the multiplication equation, and use it to write the related division fact directly to the right.

_____ | $7 \times 10 =$ _____ | | _____ $\div 7 =$ _____

_____ | $7 \times 9 =$ _____ | | _____ $\div 7 =$ _____

_____ | $7 \times 8 =$ _____ | | _____ $\div 7 =$ _____

_49_ | $7 \times 7 =$ _____ | | _____ $\div 7 =$ _____

_____ | $7 \times 6 =$ _____ | | _____ $\div 7 =$ _____

_____ | $7 \times 5 =$ _____ | | _____ $\div 7 =$ _____

_28_ | $7 \times 4 =$ _____ | | _____ $\div 7 =$ _____

_____ | $7 \times 3 =$ _____ | | _____ $\div 7 =$ _____

_____ | $7 \times 2 =$ _____ | | _____ $\div 7 =$ _____

_7_ | $7 \times 1 =$ _____ | | _____ $\div 7 =$ _____

Lesson 5:   Count by units of 7 to multiply and divide using number bonds to decompose.

EUREKA
MATH™

1. Label the tape diagram. Then, fill in the blanks below to make the statements true.

$9 \times 8 =$

$(5 \times 8) = \underline{40}$ $(\underline{4} \times 8) = 32$

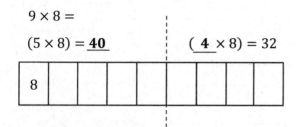

8								

$9 \times 8 = (5 + \underline{4}) \times 8$

      $= (5 \times 8) + (\underline{4} \times 8)$

      $= \quad 40 \quad + \quad \underline{32}$

      $= \quad \underline{72}$

> I can think of 9×8 as 9 eights and break apart the 9 eights into 5 eights and 4 eights. 5 eights equals 40, and 4 eights equals 32. When I add those numbers, I find that 9 eights, or 9×8, equals 72.

2. Break apart 49 to solve $49 \div 7$.

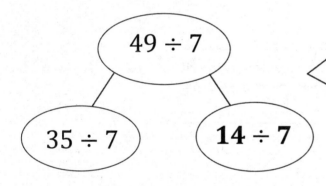

> I can use the break apart and distribute strategy to break 49 apart into 35 and 14. Those are numbers that are easier for me to divide by 7. I know that $35 \div 7 = 5$, and $14 \div 7 = 2$, so $49 \div 7$ equals $5 + 2$, which is 7.

$49 \div 7 = (35 \div 7) + (\underline{14} \div 7)$

      $= 5 + \underline{2}$

      $= \quad \underline{7}$

EUREKA MATH™

Lesson 6: Use the distributive property as a strategy to multiply and divide using units of 6 and 7.

©2018 Great Minds®. eureka-math.org

195

3. 48 third graders sit in 6 equal rows in the auditorium. How many students sit in each row? Show your thinking.

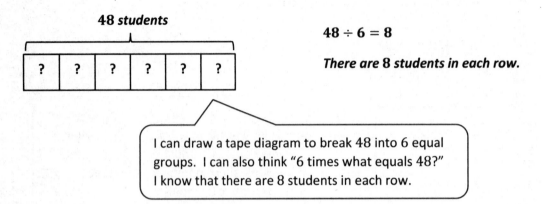

$48 \div 6 = 8$

There are 8 students in each row.

I can draw a tape diagram to break 48 into 6 equal groups. I can also think "6 times what equals 48?" I know that there are 8 students in each row.

4. Ronaldo solves 6×9 by thinking of it as $(5 \times 9) + 9$. Is he correct? Explain Ronaldo's strategy.

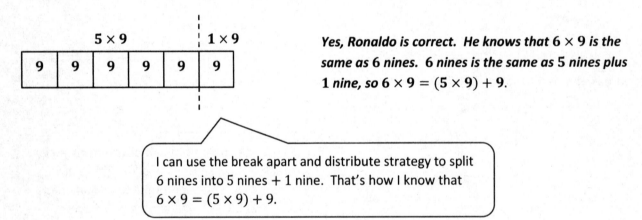

Yes, Ronaldo is correct. He knows that 6×9 is the same as 6 nines. 6 nines is the same as 5 nines plus 1 nine, so $6 \times 9 = (5 \times 9) + 9$.

I can use the break apart and distribute strategy to split 6 nines into 5 nines + 1 nine. That's how I know that $6 \times 9 = (5 \times 9) + 9$.

Lesson 6:   Use the distributive property as a strategy to multiply and divide using units of 6 and 7.

EUREKA
MATH™

Name _____   Date _____

1. Label the tape diagrams. Then, fill in the blanks below to make the statements true.

a. **6 × 7**= _____

(5 × 7) = _____   (____ × 7) = _____

(6 × 7) = (5 + 1) × 7

= (5 × 7) + (1 × 7)

= __35__ + _____

= _____

b. **7 × 7** = _____

(5 × 7) = _____   (____ × 7) = _____

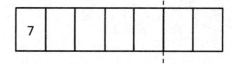

(7 × 7) = (5 + 2) × 7

= (5 × 7) + (2 × 7)

= __35__ + _____

= _____

c. **8 × 7** = _____

(5 × 7) = _____   (____ × 7) = _____

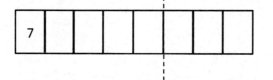

8 × 7 = (5 + _____) × 7

= (5 × 7) + (____ × 7)

= __35__ + _____

= _____

d. **9 × 7** = _____

(5 × 7) = _____   (____ × 7) = _____

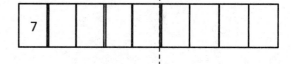

9 × 7 = (5 + _____) × 7

= (5 × 7) + (____ × 7)

= __35__ + _____

= _____

EUREKA MATH™

Lesson 6:   Use the distributive property as a strategy to multiply and divide using units of 6 and 7.

©2018 Great Minds®. eureka-math.org

197

2. Break apart 54 to solve 54 ÷ 6.

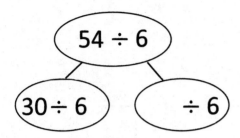

$54 ÷ 6 = (30 ÷ 6) + (\rule{2cm}{0.4pt} ÷ 6)$

$= 5 + \rule{2cm}{0.4pt}$

$= \rule{2cm}{0.4pt}$

3. Break apart 56 to solve 56 ÷ 7

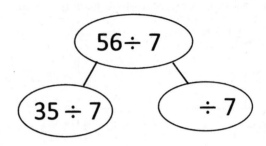

$56 ÷ 7 = (\rule{1cm}{0.4pt} ÷ \rule{1cm}{0.4pt}) + (\rule{1cm}{0.4pt} ÷ \rule{1cm}{0.4pt})$

$= 5 + \rule{1.5cm}{0.4pt}$

$= \rule{1.5cm}{0.4pt}$

4. Forty-two third grade students sit in 6 equal rows in the auditorium. How many students sit in each row? Show your thinking.

5. Ronaldo solves 7 × 6 by thinking of it as (5 × 7) + 7. Is he correct? Explain Ronaldo's strategy.

Lesson 6:   Use the distributive property as a strategy to multiply and divide using units of 6 and 7.

©2018 Great Minds®. eureka-math.org

EUREKA
MATH™

1. Match the words on the arrow to the correct equation on the target.

7 times a number equals 56

$42 \div n = 6$

The equations use n to represent the unknown number. When I read the words on the left carefully, I can pick out the correct equation on the right.

42 divided by a number equals 6

$7 \times n = 56$

2. Ari sells 7 boxes of pens at the school store.

a. Each box of pens costs $6. Draw a tape diagram, and label the total amount of money Ari makes as m dollars. Write an equation, and solve for m.

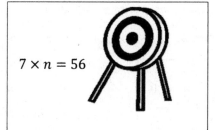

m dollars

| $6 | $6 | $6 | $6 | $6 | $6 | $6 |

$7 \times 6 = m$

$m = 42$

Ari makes $42 selling pens.

I'm using the letter m to represent how much money Ari makes. Once I find the value of m, then I know how much money Ari earns selling pens.

EUREKA MATH

Lesson 7: Interpret the unknown in multiplication and division to model and solve problems using units of 6 and 7.

199

©2018 Great Minds®. eureka-math.org

b. Each box contains 8 pens. Draw a tape diagram, and label the total number of pens as p. Write an equation, and solve for p.

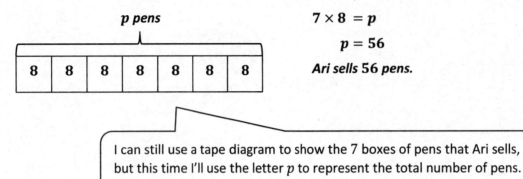

p pens

8	8	8	8	8	8	8

$7 \times 8 = p$

$p = 56$

Ari sells 56 pens.

I can still use a tape diagram to show the 7 boxes of pens that Ari sells, but this time I'll use the letter p to represent the total number of pens. Since there are 8 pens in each box, I know that the value of p is 56.

3. Mr. Lucas divides 30 students into 6 equal groups for a project. Draw a tape diagram, and label the number of students in each group as n. Write an equation, and solve for n.

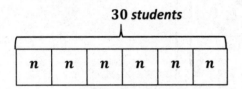

30 students

n	n	n	n	n	n

$30 \div 6 = n$

$6 \times n = 30$

$n = 5$

There are 5 students in each group.

I know that 30 students are split into 6 equal groups, so I have to solve $30 \div 6$ to figure out how many students are in each group. I'll use the letter n to represent the unknown. To solve, I can think about this as division or as an unknown factor problem.

Lesson 7:   Interpret the unknown in multiplication Iend division to model and solve problems using units of 6 and 7.

©2018 Great Minds®. eureka-math.org

EUREKA
MATH™

Name _____ Date _____

1. Match the words on the arrow to the correct equation on the target.

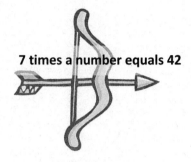

7 times a number equals 42

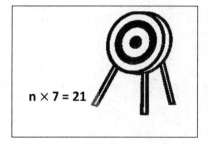

n × 7 = 21

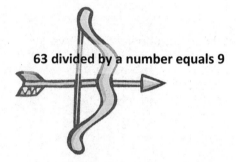

63 divided by a number equals 9

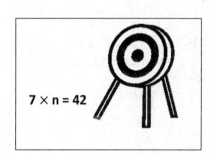

7 × n = 42

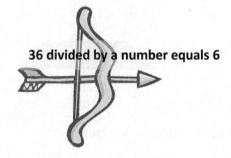

36 divided by a number equals 6

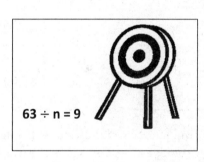

63 ÷ n = 9

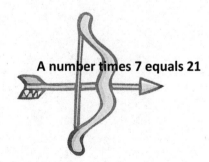

A number times 7 equals 21

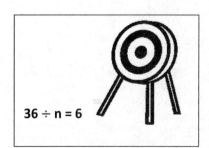

36 ÷ n = 6

EUREKA MATH

Lesson 7: Interpret the unknown in multiplication and division to model and solve problems using units of 6 and 7.

201

©2018 Great Minds®. eureka-math.org

2. Ari sells 6 boxes of pens at the school store.

   a. Each box of pens sells for $7. Draw a tape diagram, and label the total amount of money he makes as m. Write an equation, and solve for m.

   b. Each box contains 6 pens. Draw a tape diagram, and label the total number of pens as p. Write an equation, and solve for p.

3. Mr. Lucas divides 28 students into 7 equal groups for a project. Draw a tape diagram, and label the number of students in each group as n. Write an equation, and solve for n.

Lesson 7: Interpret the unknown in multiplication and division to model and solve problems using units of 6 and 7.

©2018 Great Minds®. eureka-math.org

EUREKA
MATH

1. Solve.

   a. $9 - (6 + 3) =$ ___**0**___

   > I know the parentheses mean that I have to add $6 + 3$ first. Then I can subtract that sum from 9.

   b. $(9 - 6) + 3 =$ ___**6**___

   > I know the parentheses mean that I have to subtract $9 - 6$ first. Then I can add 3. The numbers in parts (a) and (b) are the same, but the answers are different because of where the parentheses are placed.

2. Use parentheses to make the equations true.

   a. $13 = 3 + (5 \times 2)$

   > I can put parentheses around 5×2. That means I first multiply 5×2, which equals 10, and then add 3 to get 13.

   b. $16 = (3 + 5) \times 2$

   > I can put parentheses around $3 + 5$. That means I first add $3 + 5$, which equals 8, and then multiply by 2 to get 16.

3. Determine if the equation is true or false.

a. $(4 + 5) \times 2 = 18$	*True*
b. $5 = 3 + (12 \div 3)$	*False*

   > I know part (a) is true because I can add $4 + 5$, which equals 9. Then I can multiply 9×2 to get 18.

   > I know part (b) is false because I can divide 12 by 3, which equals 4. Then I can add $4 + 3$. $4 + 3$ equals 7, not 5.

EUREKA MATH™

Lesson 8: Understand the function of parentheses and apply to solving problems.

203

©2018 Great Minds®. eureka-math.org

4. Julie says that the answer to $16 + 10 - 3$ is 23 no matter where she puts the parentheses. Do you agree?

$$(16 + 10) - 3 = 23 \qquad\qquad\qquad 16 + (10 - 3) = 23$$

I agree with Julie. I put parentheses around $16 + 10$, and when I solved the equation, I got 23 because $26 - 3 = 23$. Then I moved the parentheses and put them around $10 - 3$. When I subtracted $10 - 3$ first, I still got 23 because $16 + 7 = 23$. Even though I moved the parentheses, the answer didn't change!

Lesson 8:   Understand the function of parentheses and apply to solving problems.

©2018 Great Minds®. eureka-math.org

EUREKA
MATH

Name _____ Date _____

1. Solve.

   a. 9 − (6 + 3) = _____

   b. (9 − 6) + 3 = _____

   c. _____ = 14 − (4 + 2)

   d. _____ = (14 − 4) + 2

   e. _____ = (4 + 3) × 6

   f. _____ = 4 + (3 × 6)

   g. (18 ÷ 3) + 6 = _____

   h. 18 ÷ (3 + 6) = _____

2. Use parentheses to make the equations true.

   a. 14 − 8 + 2 = 4

   b. 14 − 8 + 2 = 8

   c. 2 + 4 × 7 = 30

   d. 2 + 4 × 7 = 42

   e. 12 = 18 ÷ 3 × 2

   f. 3 = 18 ÷ 3 × 2

   g. 5 = 50 ÷ 5 × 2

   h. 20 = 50 ÷ 5 × 2

EUREKA
MATH™

Lesson 8: Understand the function of parentheses and apply to solving problems.

©2018 Great Minds®. eureka-math.org

205

3. Determine if the equation is true or false.

a. $(15 - 3) \div 2 = 6$	*Example:* True
b. $(10 - 7) \times 6 = 18$	
c. $(35 - 7) \div 4 = 8$	
d. $28 = 4 \times (20 - 13)$	
e. $35 = (22 - 8) \div 5$	

4. Jerome finds that $(3 \times 6) \div 2$ and $18 \div 2$ are equal. Explain why this is true.

5. Place parentheses in the equation below so that you solve by finding the difference between 28 and 3. Write the answer.

   $4 \times 7 - 3 =$ _____

6. Johnny says that the answer to $2 \times 6 \div 3$ is 4 no matter where he puts the parentheses. Do you agree? Place parentheses around different numbers to help you explain his thinking.

Lesson 8:   Understand the function of parentheses and apply to solving problems.

EUREKA MATH™

1. Use the array to complete the equation.

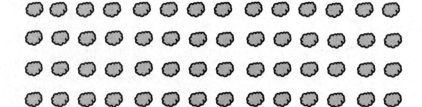

a. $4 \times 14 = $ __56__

> I can use the array to skip-count by 4 to find the product.

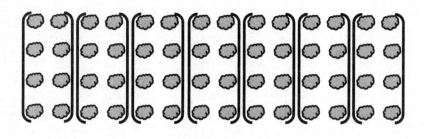

b. $(4 \times \underline{\ 2\ }) \times 7$
  $= \underline{\ 8\ } \times \underline{\ 7\ }$
  $= \underline{\ 56\ }$

> The array shows that there are 7 groups of 4×2.

> I rewrote 14 as 2×7. Then I moved the parentheses to make the equation $(4 \times 2) \times 7$. I can multiply 4×2 to get 8. Then I can multiply 8×7 to get 56. Rewriting 14 as 2×7 made the problem easier to solve!

2. Place parentheses in the equations to simplify and solve.

$3 \times 21 = 3 \times (3 \times 7)$
$\quad = (3 \times 3) \times 7$
$\quad = \underline{\ 9\ } \times 7$

$= \underline{\ 63\ }$

> I can put the parentheses around 3×3 and then multiply. 3×3 equals 9. Now I can solve the easier multiplication fact, 9×7.

EUREKA MATH

Lesson 9:   Model the associative property as a strategy to multiply.

207

©2018 Great Minds®. eureka-math.org

3. Solve. Then, match the related facts.

a. $24 \times 3 = \underline{\ 72\ } = $ $9 \times (3 \times 2)$

b. $27 \times 2 = \underline{\ 54\ } = $ $8 \times (3 \times 3)$

> I can think of 24 as 8×3. Then, I can move the parentheses to make the new expression $8 \times (3 \times 3)$. $3 \times 3 = 9$, and $8 \times 9 = 72$, so $24 \times 3 = 72$.

> I can think of 27 as 9×3. Then, I can move the parentheses to make the new expression $9 \times (3 \times 2)$. $3 \times 2 = 6$, and $9 \times 6 = 54$, so $27 \times 2 = 54$.

Lesson 9:   Model the associative property as a strategy to multiply.

EUREKA MATH™

Name _____ Date _____

1. Use the array to complete the equation.

a. 3 × 16 = _____

b. (3 × ____) × 8

= _____ × ____

= _____

c. 4 × 18 = _____

d. (4 × ____) × 9

= _____ × _____

= _____

EUREKA MATH

Lesson 9: Model the associative property as a strategy to multiply.

©2018 Great Minds®. eureka-math.org

209

2. Place parentheses in the equations to simplify and solve.

$12 \times 4 = (6 \times 2) \times 4$

$\quad = 6 \times (2 \times 4)$ $\left.\right\}$ $= \underline{\quad 48 \quad}$

$\quad = 6 \times \underline{\ 8\ }$

a. $3 \times 14 = 3 \times (2 \times 7)$

$\quad = 3 \times 2 \times 7$ $\left.\right\}$ $= \underline{\qquad}$

$\quad = \underline{\qquad} \times 7$

b. $3 \times 12 = 3 \times (3 \times 4)$

$\quad = 3 \times 3 \times 4$ $\left.\right\}$ $= \underline{\qquad}$

$\quad = \underline{\qquad} \times 4$

3. Solve. Then, match the related facts.

a. $20 \times 2 = \underline{\ 40\ } =$

b. $30 \times 2 = \underline{\qquad} =$

c. $35 \times 2 = \underline{\qquad} =$

d. $40 \times 2 = \underline{\qquad} =$

$6 \times (5 \times 2)$

$8 \times (5 \times 2)$

$4 \times (5 \times 2)$

$7 \times (5 \times 2)$

Lesson 9: Model the associative property as a strategy to multiply.

EUREKA
MATH™

1. Label the array. Then, fill in the blanks to make the statements true.

   $8 \times 6 = 6 \times 8 = $ __48__

   $(6 \times 5) = $ __30__ $(6 \times $ __3__ $) = $ __18__

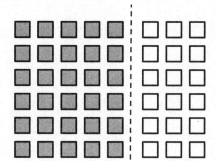

   I can use the array to help me fill in the blanks. The array shows 8 broken into 5 and 3. The shaded part shows $6 \times 5 = 30$, and the unshaded part shows $6 \times 3 = 18$. I can add the products of the smaller arrays to find the total for the entire array. $30 + 18 = 48$, so $8 \times 6 = 48$.

   $8 \times 6 = 6 \times (5 + $ __3__ $)$
   $\quad = (6 \times 5) + (6 \times $ __3__ $)$
   $\quad = \quad 30 \quad + \quad$ __18__
   $\quad = $ __48__

   The equations show the same work that I just did with the array.

2. Break apart and distribute to solve $64 \div 8$.

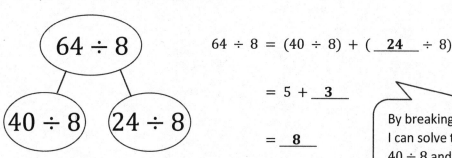

   $64 \div 8 = (40 \div 8) + ($ __24__ $\div 8)$

   $\quad = 5 + $ __3__

   $\quad = $ __8__

   By breaking 64 apart as 40 and 24, I can solve the easier division facts $40 \div 8$ and $24 \div 8$. Then I can add the quotients to solve $64 \div 8$.

   I can use a number bond instead of an array to show how to break apart $64 \div 8$.

EUREKA MATH™

Lesson 10: Use the distributive property as a strategy to multiply and divide.

211

©2018 Great Minds®. eureka-math.org

3. Count by 8. Then, match each multiplication problem with its value.

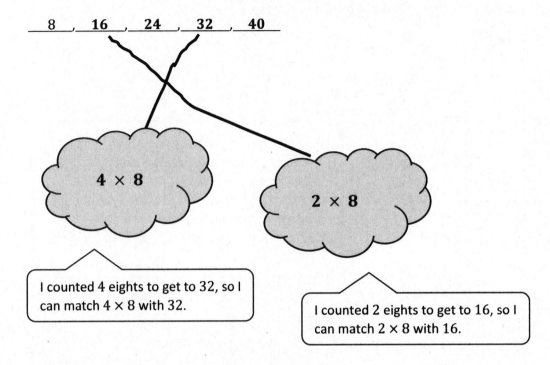

$\underline{ 8 }$, $\underline{ 16 }$, $\underline{ 24 }$, $\underline{ 32 }$, $\underline{ 40 }$

4 × 8

2 × 8

I counted 4 eights to get to 32, so I can match 4 × 8 with 32.

I counted 2 eights to get to 16, so I can match 2 × 8 with 16.

Lesson 10:   Use the distributive property as a strategy to multiply and divide.

EUREKA
MATH™

Name _____ Date _____

1. Label the array. Then, fill in the blanks to make the statements true.

$8 \times 7 = 7 \times 8 =$ _____

$(7 \times 5) =$ _____ $(7 \times$ _____$) =$ _____

$8 \times 7 = 7 \times (5 +$ _____$)$

$= (7 \times 5) + (7 \times$ _____$)$

$=$ ___35___ $+$ _____

$=$ _____

2. Break apart and distribute to solve $72 \div 8$.

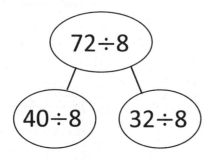

$72 \div 8 = (40 \div 8) + ($ _____ $\div 8)$

$= 5 +$ _____

$=$ _____

EUREKA MATH

Lesson 10:   Use the distributive property as a strategy to multiply and divide.

©2018 Great Minds®. eureka-math.org

213

3. Count by 8. Then, match each multiplication problem with its value.

____8____, _____, _____, _____, _____, _____, _____, _____, _____, _____

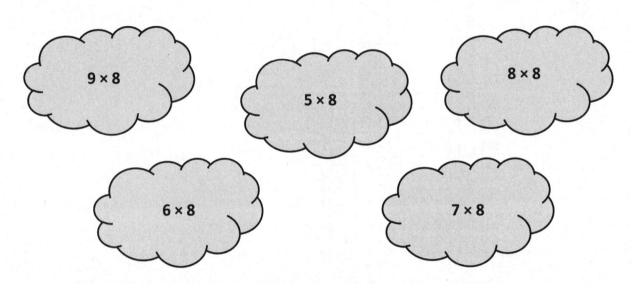

9 × 8

5 × 8

8 × 8

6 × 8

7 × 8

4. Divide.

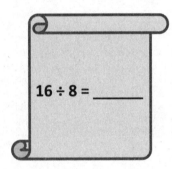

16 ÷ 8 = _____

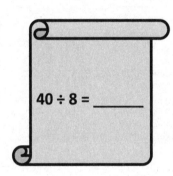

40 ÷ 8 = _____

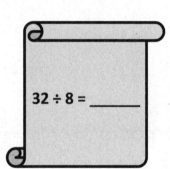

32 ÷ 8 = _____

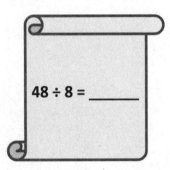

48 ÷ 8 = _____

56 ÷ 8 = _____

72 ÷ 8 = _____

Lesson 10: Use the distributive property as a strategy to multiply and divide.

EUREKA
MATH™

1. There are 8 pencils in one box. Corey buys 3 boxes. He gives an equal number of pencils to 4 friends. How many pencils does each friend receive?

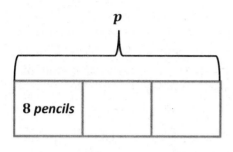

p

8 pencils

I can draw a tape diagram to help me solve. I know the number of groups is 3, and the size of each group is 8. I need to solve for the total number of pencils. I can use the letter *p* to represent the unknown.

$3 \times 8 = p$

$p = 24$

I can multiply 3×8 to find the total number of pencils Corey buys. Now I need to figure out how many pencils each friend gets.

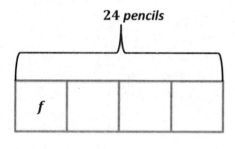

24 pencils

f

I can draw a tape diagram with 4 units to represent the 4 friends. I know that the total is 24 pencils. I need to solve for the size of each group. I can use the letter *f* to represent the unknown.

$24 \div 4 = f$

$f = 6$

I can divide 24 by 4 to find the number of pencils each friend gets.

Each friend receives 6 pencils.

EUREKA MATH™

Lesson 11: Interpret the unknown in multiplication and division to model and solve problems.

©2018 Great Minds®. eureka-math.org

215

2. Lilly makes $7 each hour she babysits. She babysits for 8 hours. Lilly uses her babysitting money to buy a toy. After buying the toy, she has $39 left. How much money did Lilly spend on the toy?

b

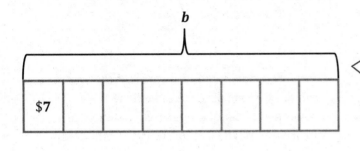

$7

> I can draw a tape diagram to help me solve. I know the number of groups is 8, and the size of each group is $7. I need to solve for the total amount of money. I can use the letter *b* to represent the unknown.

$8 \times \$7 = b$

$b = \$56$

> I can multiply $8 \times \$7$ to find the total amount of money Lilly earns babysitting. Now I need to figure out how much money she spent on the toy.

$56

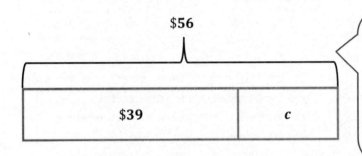

$39   *c*

> I can draw a tape diagram with two parts and a total of $56. One part represents the amount of money Lilly has left, $39. The other part is the unknown and represents the amount of money Lilly spent on the toy. I can use the letter *c* to represent the unknown.

$\$56 - \$39 = c$

> I can subtract $\$56 - \$39$ to find the amount of money Lilly spent on the toy.

$\$57 - \$40 = \$17$

> I can use compensation to subtract using mental math. I do that by adding 1 to each number, which makes it easier for me to solve.

$c = \$17$

$$\begin{array}{r} {}^{4}\;{}^{16} \\ \$\;\not{5}\,\not{6} \\ -\$\;3\;9 \\ \hline \$\;1\;7 \end{array}$$

> Or I can use the standard algorithm for subtraction.

Lilly spent $17 on the toy.

Lesson 11:   Interpret the unknown in multiplication and division to model and solve problems.

©2018 Great Minds®. eureka-math.org

EUREKA MATH

Name _____   Date _____

1. Jenny bakes 10 cookies. She puts 7 chocolate chips on each cookie. Draw a tape diagram, and label the total amount of chocolate chips as *c*. Write an equation, and solve for *c*.

_____

2. Mr. Lopez arranges 48 dry erase markers into 8 equal groups for his math stations. Draw a tape diagram, and label the number of dry erase markers in each group as *v*. Write an equation, and solve for *v*.

_____

3. There are 35 computers in the lab. Five students each turn off an equal number of computers. How many computers does each student turn off? Label the unknown as *m*, and then solve.

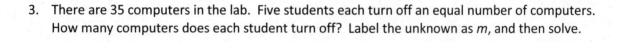

EUREKA MATH   **Lesson 11:**   Interpret the unknown in multiplication and division to model and solve problems.   **217**

©2018 Great Minds®. eureka-math.org

4. There are 9 bins of books. Each bin has 6 comic books. How many comic books are there altogether?

_____

5. There are 8 trail mix bags in one box. Clarissa buys 5 boxes. She gives an equal number of bags of trail mix to 4 friends. How many bags of trail mix does each friend receive?

_____

6. Leo earns $8 each week for doing chores. After 7 weeks, he buys a gift and has $38 left. How much money does he spend on the gift?

Lesson 11: Interpret the unknown in multiplication and division to model and solve problems.

©2018 Great Minds®. eureka-math.org

EUREKA MATH™

1. Each 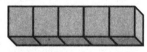 has a value of 9. Find the value of each row. Then, add the rows to find the total.

$7 \times 9 =$ __63__

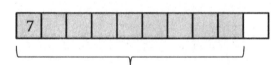

$5 \times 9 = 45$

__2__ $\times 9 =$ __18__

> I know each cube has a value of 9. The 2 rows of cubes show 7 nines broken up as 5 nines and 2 nines. It is the break apart and distribute strategy using the familiar fives fact.

$$7 \times 9 = (5 + \underline{\ 2\ }) \times 9$$
$$= (5 \times 9) + (\underline{\ 2\ } \times 9)$$
$$= 45 + \underline{\ 18\ }$$
$$= \underline{\ 63\ }$$

> To add 45 and 18, I'll simplify by taking 2 from 45. I'll add the 2 to 18 to make 20. Then I can think of the problem as $43 + 20$.

2. Find the total value of the shaded blocks.

$9 \times 7 =$

7								

9 sevens = 10 sevens − 1 seven

$= \underline{\ 70\ } -7$

$= \underline{\ 63\ }$

> This shows a different way to solve. I can think of 7 nines as 9 sevens. 9 is closer to 10 than it is to 5. So instead of using a fives fact, I can use a tens fact to solve. I take the product of 10 sevens and subtract 1 seven.

> This strategy made the math simpler and more efficient. I can do $70 - 7$ quickly in my head!

3. James buys a pack of baseball cards. He counts 9 rows of 6 cards. He thinks of 10 sixes to find the total number of cards. Show the strategy that James might have used to find the total number of baseball cards.

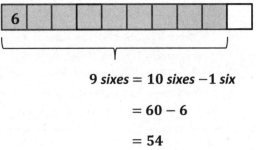

$$9\ sixes = 10\ sixes - 1\ six$$

$$= 60 - 6$$

$$= 54$$

James bought 54 baseball cards.

James uses the tens fact to solve for the nines fact. To solve for 9 sixes, he starts with 10 sixes and subtracts 1 six.

Lesson 12:   Apply the distributive property and the fact $9 = 10 - 1$ as a strategy to multiply.

©2018 Great Minds®. eureka-math.org

EUREKA MATH™

Name _____   Date _____

1. Find the value of each row. Then, add the rows to find the total.

a. Each  has a value of 6.

9 × 6 = _____

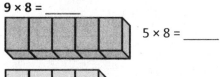

5 × 6 = 30

4 × 6 = _____

9 × 6 = (5 + 4) × 6
= (5 × 6) + (4 × 6)
= 30 + _____
= _____

b. Each has a value of 7.

9 × 7 = _____

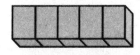

5 × 7 = _____

_____ × 7 = _____

9 × 7 = (5 + _____) × 7
= (5 × 7) + (_____ × 7)
= 35 + _____
= _____

c. Each has a value of 8.

9 × 8 = _____

5 × 8 = _____

_____ × 8 = _____

9 × 8 = (5 + _____) × 8
= (5 × 8) + (_____ × _____)
= 40 + _____
= _____

d. Each 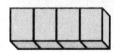 has a value of 9.

9 × 9 = _____

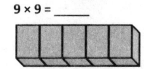

5 × 9 = _____

_____ × 9 = _____

9 × 9 = (5 + _____) × 9
= (5 × 9) + (_____ × _____)
= 45 + _____
= _____

EUREKA MATH™   **Lesson 12:**   Apply the distributive property and the fact $9 = 10 - 1$ as a strategy to multiply.   **221**

©2018 Great Minds®. eureka-math.org

2. Match.

a. **9 fives** = 10 fives − 1 five

      = 50 − 5

b. **9 sixes** = 10 sixes − 1 six

     = _____ − 6

c. **9 sevens** = 10 sevens − 1 seven

     = _____ − 7

d. **9 eights** = 10 eights − 1 eight

     = _____ − 8

e. **9 nines** = 10 nines − 1 nine

     = _____ − _____

f. **9 fours** = 10 fours − 1 four

     = _____ − _____

Lesson 12:   Apply the distributive property and the fact $9 = 10 − 1$ as a strategy to multiply.

EUREKA MATH™

1. Complete to make true statements.

   a. 10 more than 0 is __**10**__,

   1 less is __**9**__.

   $1 \times 9 =$ __**9**__

   > These statements show a simplifying strategy for skip-counting by nine. It's a pattern of adding 10 and then subtracting 1.

   b. 10 more than 9 is __**19**__,

   1 less is __**18**__.

   $2 \times 9 =$ __**18**__

   > I notice another pattern! I compare the digits in the ones and tens places of the multiples. I can see that from one multiple to the next, the digit in the tens place increases by 1, and the digit in the ones place decreases by 1.

   c. 10 more than 18 is __**28**__,

   1 less is __**27**__.

   $3 \times 9 =$ __**27**__

2.

   a. Analyze the skip-counting strategy in Problem 1. What is the pattern?

   The pattern is add 10 and then subtract 1.

   To get a nines fact, you add 10 and then subtract 1.

   b. Use the pattern to find the next 2 facts. Show your work.

$4 \times 9 =$	$27 + 10 = 37$	$5 \times 9 =$	$36 + 10 = 46$
	$37 - 1 = 36$		$46 - 1 = 45$
	$4 \times 9 = 36$		$5 \times 9 = 45$

   > I can check my answers by adding the digits of each multiple. I know that multiples of 9 I've learned have a sum of digits equal to 9. If the sum isn't equal to 9, I've made a mistake. I know 36 is correct because $3 + 6 = 9$. I know 45 is correct because $4 + 5 = 9$.

EUREKA MATH™

Lesson 13: Identify and use arithmetic patterns to multiply.

Name _____   Date _____

1. a. Skip-count by nines down from 90.

____90___, _____, ___72___, _____, _____, _____, ___36___, _____, _____, _____

   b. Look at the *tens* place in the count-by. What is the pattern?

   c. Look at the *ones* place in the count-by. What is the pattern?

2. Each equation contains a letter representing the unknown. Find the value of each unknown.

$a \times 9 = 18$

$a =$ _____

$m \div 9 = 3$

$m =$ _____

$e \times 9 = 45$

$e =$ _____

$f \div 9 = 4$

$f =$ _____

$9 \times d = 81$

$d =$ _____

$w \div 9 = 6$

$w =$ _____

$9 \times s = 90$

$s =$ _____

$k \div 9 = 8$

$k =$ _____

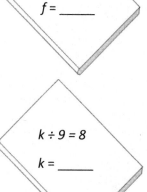

3. Solve.

a. What is 10 more than 0? ____   b. What is 10 more than 9? ____   c. What is 10 more than 18? ____

What is 1 less? ____   What is 1 less? ____   What is 1 less? ____

$1 \times 9 =$ ____ $2 \times 9 =$ ____ $3 \times 9 =$ ____

d. What is 10 more than 27? ____   e. What is 10 more than 36? ____   f. What is 10 more than 45? ____

What is 1 less? ____   What is 1 less? ____   What is 1 less? ____

$4 \times 9 =$ ____ $5 \times 9 =$ ____ $6 \times 9 =$ ____

g. What is 10 more than 54? ____   h. What is 10 more than 63? ____   i. What is 10 more than 72? ____

What is 1 less? ____   What is 1 less? ____   What is 1 less? ____

$7 \times 9 =$ ____ $8 \times 9 =$ ____ $9 \times 9 =$ ____

j. What is 10 more than 81? ____

What is 1 less? ____

$10 \times 9 =$ ____

4. Explain the pattern in Problem 3, and use the pattern to solve the next 3 facts.

$11 \times 9 =$ ____ $12 \times 9 =$ ____ $13 \times 9 =$ ____

Lesson 13:   Identify and use arithmetic patterns to multiply.

EUREKA MATH

1. Tracy figures out the answer to 7×9 by putting down her right index finger (shown). What is the answer? Explain how to use Tracy's finger strategy.

In order for this strategy to work, I have to imagine that my fingers are numbered 1 through 10, with my pinky on the left being number 1 and my pinky on the right being number 10.

Tracy first lowers the finger that matches the number of nines, 7. She sees that there are 6 fingers to the left of the lowered finger, which is the digit in the tens place, and that there are 3 fingers to the right of the lowered finger, which is the digit in the ones place. So, Tracy's fingers show that the product of 7×9 is 63.

2. Chris writes $54 = 9 \times 6$. Is he correct? Explain 3 strategies Chris can use to check his work.

Chris can use the $9 = 10 - 1$ strategy to check his answer.

$$9 \times 6 = (10 \times 6) - (1 \times 6)$$
$$= 60 - 6$$
$$= 54$$

He can also check his answer by finding the sum of the digits in the product to see if it equals 9. Since $5 + 4 = 9$, his answer is correct.

A third strategy for checking his answer is to use the number of groups, 6, to find the digits in the tens place and ones place of the product. He can use $6 - 1 = 5$ to get the digit in the tens place, and $10 - 6 = 4$ to get the digit in the ones place. This strategy also shows that Chris's answer is correct.

Chris can also use the add 10, subtract 1 strategy to list all the nines facts, or he can use the break apart and distribute strategy with fives facts. For example, he can think of 9 sixes as 5 sixes + 4 sixes. There are many strategies and patterns that can help Chris check his work multiplying with nine.

Name _____ Date _____

1. a.   Multiply. Then, add the digits in each product.

$10 \times 9 = 90$	___9___ + ___0___ = ___9___
$9 \times 9 = 81$	___8___ + ___1___ = ___9___
$8 \times 9 =$	_____ + _____ = _____
$7 \times 9 =$	_____ + _____ = _____
$6 \times 9 =$	_____ + _____ = _____
$5 \times 9 =$	_____ + _____ = _____
$4 \times 9 =$	_____ + _____ = _____
$3 \times 9 =$	_____ + _____ = _____
$2 \times 9 =$	_____ + _____ = _____
$1 \times 9 =$	_____ + _____ = _____

   b.   What pattern did you notice in Problem 1(a)? How can this strategy help you check your work with nines facts?

2. Thomas calculates 9 × 7 by thinking about it as 70 – 7 = 63. Explain Thomas' strategy.

_____

3. Alexia figures out the answer to 6 × 9 by lowering the thumb on her right hand (shown). What is the answer? Explain Alexia's strategy.

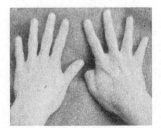

_____

4. Travis writes 72 = 9 × 8. Is he correct? Explain at least 2 strategies Travis can use to check his work.

Lesson 14:   Identify and use arithmetic patterns to multiply.

Judy wants to give each of her friends a bag of 9 marbles. She has a total of 54 marbles. She runs to give them to her friends and gets so excited that she drops and loses 2 bags. How many total marbles does she have left to give away?

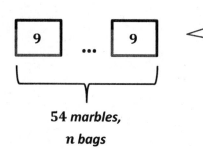

54 marbles,

n bags

I can model the problem using a tape diagram. I know Judy has a total of 54 marbles, and each bag has 9 marbles. I don't know how many bags of marbles Judy has at first. Since I know the size of each group is 9 but I don't know the number of groups, I put a "…" in between the 2 units to show that I don't yet know how many groups, or units, to draw.

n represents the number of bags of marbles

$54 \div 9 = n$

$n = 6$

I can use the letter n to represent the unknown, which is the number of bags Judy has at first. I can find the unknown by dividing 54 by 9 to get 6 bags. But 6 bags does not answer the question, so my work on this problem is not finished.

Now I can redraw my model to show the 6 bags of marbles. I know that Judy drops and loses 2 bags. The unknown is the total number of marbles she has left to give away. I can represent this unknown with the letter m.

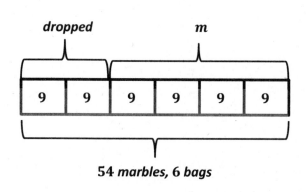

54 marbles, 6 bags

m represents the total number of marbles left

$4 \times 9 = m$

$m = 36$

Judy still has 36 marbles left to give away.

From my diagram, I can see that Judy has 4 bags of 9 marbles left. I can choose any of my nines strategies to help me solve 4×9. $4 \times 9 = 36$, which means there are 36 total marbles left.

I read the problem carefully and made sure to answer with the total number of marbles, not the number of bags. Putting my answer in a statement helps me check that I've answered the problem correctly.

EUREKA MATH™

Lesson 15:   Interpret the unknown in multiplication and division to model and solve problems.

©2018 Great Minds®. eureka-math.org

231

Name _____ Date _____

1. The store clerk equally divides 36 apples among 9 baskets. Draw a tape diagram, and label the number of apples in each basket as *a*. Write an equation, and solve for *a*.

2. Elijah gives each of his friends a pack of 9 almonds. He gives away a total of 45 almonds. How many packs of almonds did he give away? Model using a letter to represent the unknown, and then solve.

3. Denice buys 7 movies. Each movie costs $9. What is the total cost of 7 movies? Use a letter to represent the unknown. Solve.

Lesson 15: Interpret the unknown in multiplication and division to model and solve problems.

©2018 Great Minds®. eureka-math.org

233

4. Mr. Doyle shares 1 roll of bulletin board paper equally with 8 teachers. The total length of the roll is 72 meters. How much bulletin board paper does each teacher get?

5. There are 9 pens in a pack. Ms. Ochoa buys 9 packs. After giving her students some pens, she has 27 pens left. How many pens did she give away?

6. Allen buys 9 packs of trading cards. There are 10 cards in each pack. He can trade 30 cards for a comic book. How many comic books can he get if he trades all of his cards?

Lesson 15:   Interpret the unknown in multiplication and division to model and solve problems.

©2018 Great Minds®. eureka-math.org

EUREKA
MATH™

1. Let $g = 4$. Determine whether the equations are true or false.

a. $g \times 0 = 0$	**True**
b. $0 \div g = 4$	**False**
c. $1 \times g = 1$	**False**
d. $g \div 1 = 4$	**True**

I know this equation is false because 0 divided by any number is 0. If I put in any value for g other than 0, the answer will be 0.

I know this is false because any number times 1 equals that number, not 1. This equation would be correct if it was written as $1 \times g = 4$.

2. Elijah says that any number multiplied by 1 equals that number.

   a. Write a multiplication equation using c to represent Elijah's statement.

   $1 \times c = c$

   I can also use the commutative property to write my equation as $c \times 1 = c$.

   b. Using your equation from part (a), let $c = 6$, and draw a picture to show that the new equation is true.

   My picture shows 1 group multiplied by c, or 6. 1 group of 6 makes a total of 6. This works for any value, not just 6.

EUREKA
MATH™

Lesson 16:   Reason about and explain arithmetic patterns using units of 0 and 1 as they relate to multiplication and division.

©2018 Great Minds®. eureka-math.org

235

Name _____ Date _____

1. Complete.

   a. $4 \times 1 =$ _____

   b. $4 \times 0 =$ _____

   c. _____ $\times 1 = 5$

   d. _____ $\div 5 = 0$

   e. $6 \times$ _____ $= 6$

   f. _____ $\div 6 = 0$

   g. $0 \div 7 =$ _____

   h. $7 \times$ _____ $= 0$

   i. $8 \div$ _____ $= 8$

   j. _____ $\times 8 = 8$

   k. $9 \times$ _____ $= 9$

   l. $9 \div$ _____ $= 1$

2. Match each equation with its solution.

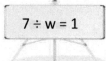

$9 \times 1 = w$

$w \times 1 = 6$

$7 \div w = 1$

$1 \times w = 8$

$w \div 8 = 0$

$9 \div 9 = w$

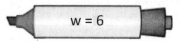

$w = 6$

$w = 7$

$w = 8$

$w = 9$

$w = 1$

$w = 0$

Lesson 16: Reason about and explain arithmetic patterns using units of 0 and 1 as they relate to multiplication and division.

237

EUREKA MATH™

©2018 Great Minds®. eureka-math.org

3. Let $c = 8$. Determine whether the equations are true or false. The first one has been done for you.

a. $c \times 0 = 8$	False
b. $0 \times c = 0$	
c. $c \times 1 = 8$	
d. $1 \times c = 8$	
e. $0 \div c = 8$	
f. $8 \div c = 1$	
g. $0 \div c = 0$	
h. $c \div 0 = 8$	

4. Rajan says that any number multiplied by 1 equals that number.

a. Write a multiplication equation using n to represent Rajan's statement.

b. Using your equation from Part (a), let $n = 5$, and draw a picture to show that the new equation is true.

Lesson 16:   Reason about and explain arithmetic patterns using units of 0 and 1 as
they relate to multiplication and division.

©2018 Great Minds®. eureka-math.org

EUREKA
MATH

1. Explain how $8 \times 7 = (5 \times 7) + (3 \times 7)$ is shown in the multiplication table.

    The multiplication table shows $5 \times 7 = 35$ and $3 \times 7 = 21$. So, $35 + 21 = 56$, which is the product of 8×7.

    This is the break apart and distribute strategy. Using that strategy, I can add the products of 2 smaller facts to find the product of a larger fact.

2. Use what you know to find the product of 3×16.

    $$3 \times 16 = (3 \times 8) + (3 \times 8)$$
    $$= 24 + 24$$
    $$= 48$$

    I can also break up 3×16 as 10 threes + 6 threes, which is $30 + 18$. Or I can add 16 three times: $16 + 16 + 16$. I always want to use the most efficient strategy. This time it helped me to see the problem as double 24.

3. Today in class we found that $n \times n$ is the sum of the first n odd numbers. Use this pattern to find the value of n for each equation below.

    a. $1 + 3 + 5 = n \times n$

        $9 = 3 \times 3$

    b. $1 + 3 + 5 + 7 = n \times n$

        $16 = 4 \times 4$

    The sum of the first 3 odd numbers is the same as the product of 3×3. The sum of the first 4 odd numbers is the same as the product of 4×4. The sum of the first 5 odd numbers is the same as the product of 5×5.

    c. $1 + 3 + 5 + 7 + 9 = n \times n$

        $25 = 5 \times 5$

    Wow, it's a pattern! I know that the first 6 odd numbers will be the same as the product of 6×6 and so on.

Name _____ Date _____

1. a.   Write the products into the chart as fast as you can.

×	1	2	3	4	5	6	7	8
1								
2								
3								
4								
5								
6								
7								
8								

   b.   Color the rows and columns with even factors yellow.

   c.   What do you notice about the factors and products that are left unshaded?

EUREKA MATH

Lesson 17:   Identify patterns in multiplication and division facts using the
                multiplication table.

241

©2018 Great Minds®. eureka-math.org

d. Complete the chart by filling in each blank and writing an example for each rule.

Rule	Example
odd times odd equals _____	
even times even equals _____	
even times odd equals _____	

e. Explain how $7 \times 6 = (5 \times 6) + (2 \times 6)$ is shown in the table.

f. Use what you know to find the product of 4×16 or 8 fours + 8 fours.

2. Today in class, we found that $n \times n$ is the sum of the first n odd numbers. Use this pattern to find the value of n for each equation below. The first is done for you.

a. $1 + 3 + 5 = n \times n$

$9 = 3 \times 3$

b. $1 + 3 + 5 + 7 = n \times n$

Lesson 17:   Identify patterns in multiplication and division facts using the
multiplication table.

©2018 Great Minds®. eureka-math.org

EUREKA
MATH™

c. $1 + 3 + 5 + 7 + 9 + 11 = n \times n$

d. $1 + 3 + 5 + 7 + 9 + 11 + 13 + 15 = n \times n$

e. $1 + 3 + 5 + 7 + 9 + 11 + 13 + 15 + 17 + 19 = n \times n$

EUREKA
MATH™

Lesson 17: Identify patterns in multiplication and division facts using the multiplication table.

243

©2018 Great Minds®. eureka-math.org

William has $187 in the bank. He saves the same amount of money each week for 6 weeks and puts it in the bank. Now William has $241 in the bank. How much money does William save each week?

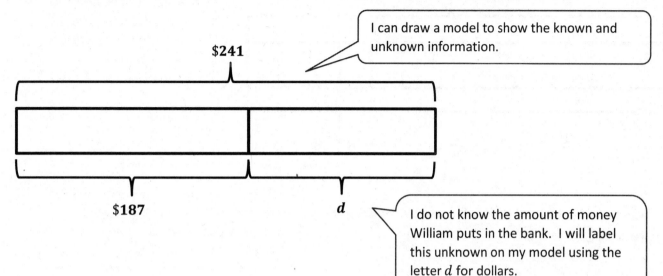

$241

I can draw a model to show the known and unknown information.

$187   d

I do not know the amount of money William puts in the bank. I will label this unknown on my model using the letter d for dollars.

d represents the number of dollars William puts in the bank

$241 − $187 = d

d = $54

I can write what d represents and then write an equation to solve for d. I can subtract the known part, $187, from the whole amount, $241, to find d.

This answer is reasonable because $187 + $54 = $241. But it does not answer the question the problem asks. I'm trying to figure out how much money William saves each week, so I need to adjust my model.

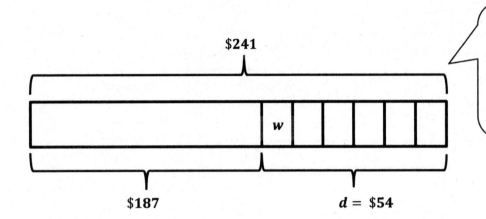

I can split the $54 into 6 equal parts to show the 6 weeks. I label the unknown *w* to represent how much money William saves each week.

w represents the number of dollars saved each week

$54 ÷ 6 = w

w = $9

I will write what *w* represents and then write an equation to solve for *w*. I can divide $54 by 6 to get $9.

William saves $9 each week.

My answer is reasonable because $9 a week for 6 weeks is $54. That's about $50. $187 is about $190. $190 + $50 = $240, which is very close to $241. My estimate is only $1 less than my answer!

I can explain why my answer is reasonable by estimating.

      Lesson 18:   Solve two-step word problems involving all four operations and assess the reasonableness of solutions.

©2018 Great Minds®. eureka-math.org

EUREKA
MATH™

Name _____   Date _____

Use the RDW process for each problem. Explain why your answer is reasonable.

1. Mrs. Portillo's cat weighs 6 kilograms. Her dog weighs 22 kilograms more than her cat. What is the total weight of her cat and dog?

2. Darren spends 39 minutes studying for his science test. He then does 6 chores. Each chore takes him 3 minutes. How many minutes does Darren spend studying and doing chores?

3. Mr. Abbot buys 8 boxes of granola bars for a party. Each box has 9 granola bars. After the party, there are 39 bars left. How many bars were eaten during the party?

EUREKA MATH™

Lesson 18:   Solve two-step word problems involving all four operations and assess
                  the reasonableness of solutions.

©2018 Great Minds®. eureka-math.org

247

4. Leslie weighs her marbles in a jar, and the scale reads 474 grams. The empty jar weighs 439 grams. Each marble weighs 5 grams. How many marbles are in the jar?

5. Sharon uses 72 centimeters of ribbon to wrap gifts. She uses 24 centimeters of her total ribbon to wrap a big gift. She uses the remaining ribbon for 6 small gifts. How much ribbon will she use for each small gift if she uses the same amount on each?

6. Six friends equally share the cost of a gift. They pay $90 and receive $42 in change. How much does each friend pay?

Lesson 18: Solve two-step word problems involving all four operations and assess the reasonableness of solutions.

©2018 Great Minds®. eureka-math.org

EUREKA
MATH™

1. Use the disks to fill in the blanks in the equations.

This array of disks shows 2 rows of 3 ones.

This array of disks shows 2 rows of 3 tens.

a.

2 × 3 ones = ____6____ ones

2 × 3 = ___6___

b.

2 × 3 tens = ____6____ tens

2 × 30 = ___60___

The top equations are written in unit form. The bottom equations are written in standard form. The 2 equations say the same thing.

I see that both arrays have the same number of disks. The only difference is the unit. The array on the left uses ones, and the array on the right uses tens.

EUREKA MATH

Lesson 19:   Multiply by multiples of ten using the place value chart.

249

©2018 Great Minds®. eureka-math.org

> I see that the difference between Problems 1 and 2 is the model. Problem 1 uses place value disks. Problem 2 uses the chip model. With both models, I'm still multiplying ones and tens.

2. Use the chart to complete the blanks in the equations.

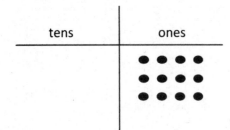

tens	ones	tens	ones

a. 3 × 4 ones = __12__ ones

    3 × 4 = __12__

b. 3 × 4 tens = __12__ tens

    3 × 40 = __120__

> I notice the number of dots is exactly the same in both charts. The difference between the charts is that when the units change from ones to tens, the dots shift over to the tens place.

3. Match.

    80 × 2   160

> In order to solve a more complicated problem like this one, I can first think of it as 8 ones × 2, which is 16. Then all I need to do is move the answer over to the tens place so it becomes 16 tens. 16 tens is the same as 160.

EUREKA
MATH™

Name _____   Date _____

1. Use the disks to complete the blanks in the equations.

a.

3 × 3 ones = _____ ones

3 × 3 = _____

b.

3 × 3 tens = _____ tens

30 × 3 = _____

2. Use the chart to complete the blanks in the equations.

tens	ones
	● ● ● ● ● ● ● ● ● ●

tens	ones
● ● ● ● ● ● ● ● ● ●	

a.   2 × 5 ones = _____ ones

     2 × 5 = _____

b.   2 × 5 tens = _____ tens

     2 × 50 = _____

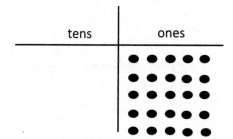

tens	ones
	● ● ● ● ● ● ● ● ● ● ● ● ● ● ● ● ● ● ● ● ● ● ● ● ●

tens	ones
● ● ● ● ● ● ● ● ● ● ● ● ● ● ● ● ● ● ● ● ● ● ● ● ●	

c.   5 × 5 ones = _____ ones

     5 × 5 = _____

d.   5 × 5 tens = _____ tens

     5 × 50 = _____

3. Match.

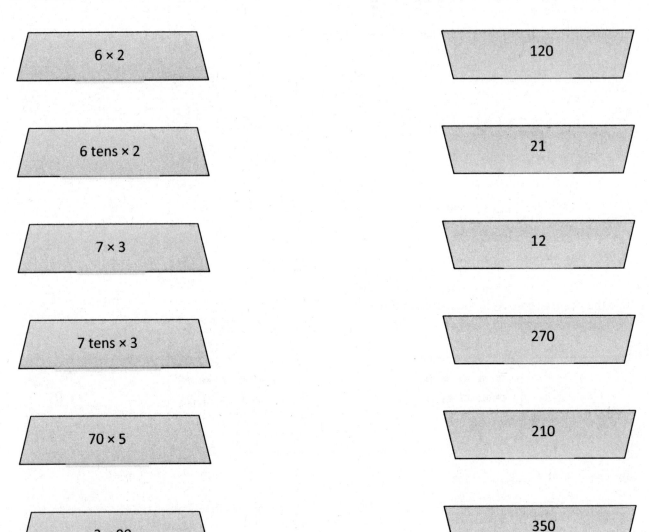

6 × 2		120
6 tens × 2		21
7 × 3		12
7 tens × 3		270
70 × 5		210
3 × 90		350

4. Each classroom has 30 desks. What is the total number of desks in 8 classrooms? Model with a tape diagram.

Lesson 19:   Multiply by multiples of ten using the place value chart.

EUREKA
MATH™

1. Use the chart to complete the equations. Then solve.

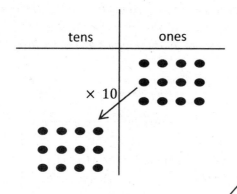

I know that parentheses change the way numbers are grouped for solving. I can see that the parentheses group 3 × 4 ones, so I'll do that part of the equation first. 3 × 4 ones = 12 ones. Next I'll multiply the 12 ones by 10. The equation becomes 12 × 10 = 120. The chip model shows how I can multiply the 3 groups of 4 ones by 10.

a.   (3 × 4) × 10

= (12 ones) × 10

= __120__

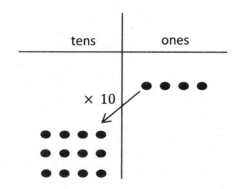

I can see that here the parentheses move over and group the 4 ones × 10. I'll solve that first to get 40, or 4 tens. Then I can multiply the 4 tens by 3. So the equation becomes 3 × 40 = 120. The chip model shows how I multiply 4 ones by 10 first and then multiply the 4 tens by three.

b.   3 × (4 × 10)

= 3 × (4 tens)

= __120__

By moving the parentheses over and grouping the numbers differently, this problem becomes friendlier. 3 × 40 is a little easier than multiplying 12 × 10. This new strategy will me help find larger unknown facts later on.

EUREKA MATH™

Lesson 20:   Use place value strategies and the associative property $n × (m × 10) = (n × m) × 10$ (where n and m are less than 10) to multiply multiples of 10.

253

©2018 Great Minds®. eureka-math.org

2. John solves 30×3 by thinking about 10×9. Explain his strategy.

$$30 \times 3 = (10 \times 3) \times 3$$
$$= 10 \times (3 \times 3)$$
$$= 10 \times 9$$
$$= 90$$

John writes 30×3 as $(10 \times 3) \times 3$. Then he moves the parentheses over to group 3×3. Solving 3×3 first makes the problem easier. Instead of 30×3, John can solve by thinking of an easier fact, 10×9.

Although it is easy to solve for 30×3, John moves the parentheses over and groups the numbers differently to make the problem a little friendlier for him. It's just another way to think about the problem.

Lesson 20: Use place value strategies and the associative property $n \times (m \times 10) =$ $(n \times m) \times 10$ (where n and m are less than 10) to multiply multiples of 10.

EUREKA MATH

Name _____ Date _____

1. Use the chart to complete the equations. Then, solve.

tens	ones

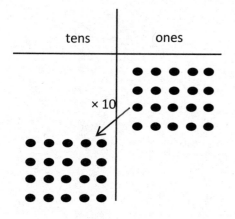

a. $(2 \times 5) \times 10$

= (10 ones) × 10

= _____

tens	ones

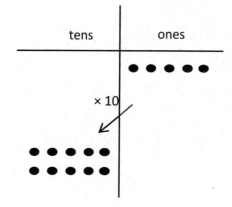

b. $2 \times (5 \times 10)$

= 2 × (5 tens)

= _____

tens	ones

c. $(4 \times 5) \times 10$

= (_____ ones) × 10

= _____

tens	ones

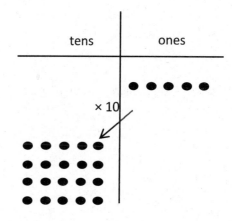

d. $4 \times (5 \times 10)$

= 4 × (_____ tens)

= _____

EUREKA MATH

Lesson 20: Use place value strategies and the associative property $n \times (m \times 10) = (n \times m) \times 10$ (where n and m are less than 10) to multiply multiples of 10.

255

©2018 Great Minds®. eureka-math.org

2.   Solve. Place parentheses in (c) and (d) as needed to find the related fact.

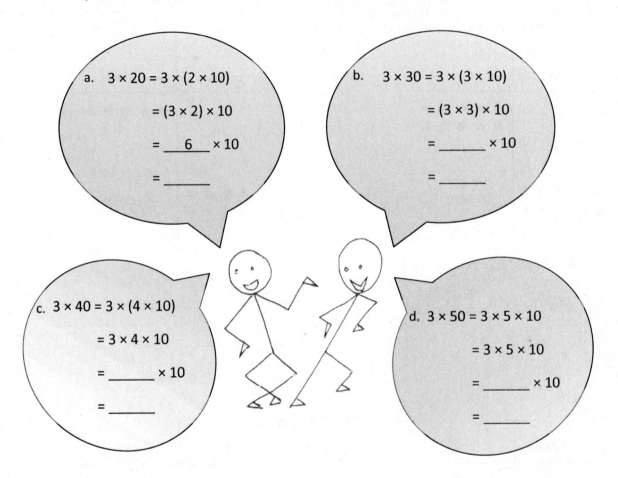

a. $3 \times 20 = 3 \times (2 \times 10)$

$= (3 \times 2) \times 10$

$= \underline{6} \times 10$

$= \underline{}$

b. $3 \times 30 = 3 \times (3 \times 10)$

$= (3 \times 3) \times 10$

$= \underline{} \times 10$

$= \underline{}$

c. $3 \times 40 = 3 \times (4 \times 10)$

$= 3 \times 4 \times 10$

$= \underline{} \times 10$

$= \underline{}$

d. $3 \times 50 = 3 \times 5 \times 10$

$= 3 \times 5 \times 10$

$= \underline{} \times 10$

$= \underline{}$

3.   Danny solves 5×20 by thinking about 10×10. Explain his strategy.

Lesson 20:   Use place value strategies and the associative property $n \times (m \times 10) = (n \times m) \times 10$ (where n and m are less than 10) to multiply multiples of 10.

©2018 Great Minds®. eureka-math.org

EUREKA
MATH™

Jen makes 34 bracelets. She gives 19 bracelets away as gifts and sells the rest for $3 each. She would like to buy an art set that costs $55 with the money she earns. Does she have enough money to buy it? Explain why or why not.

> I can draw a model to show the known and unknown information. I can see from my drawing that I need to find a missing part. I can label my missing part with a *b* to represent the number of bracelets Jen has left to sell.

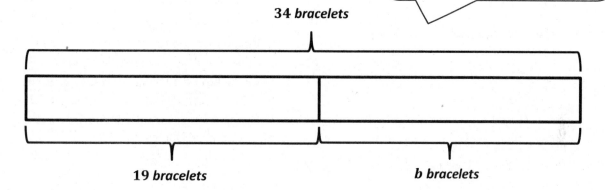

34 bracelets

19 bracelets   **b bracelets**

b represents the number of bracelets Jen has left to sell

$$34 - 19 = b$$
$$b = 15$$

> I can write what *b* represents and then write an equation to solve for *b*. I subtract the given part, 19, from the whole amount, 34. I can use a compensation strategy to think of $34 - 19$ as $35 - 20$ because $35 - 20$ is an easier fact to solve. Jen has 15 bracelets left.

> This answer is reasonable because $19 + 15 = 34$. But it doesn't answer the question in the problem. Next, I have to figure out how much money Jen earns from selling the 15 bracelets, so I need to adjust my model.

EUREKA MATH™

Lesson 21: Solve two-step word problems involving multiplying single-digit factors and multiplies of 10.

©2018 Great Minds®. eureka-math.org

257

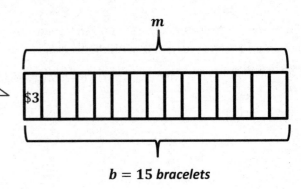

m

$3

b = 15 *bracelets*

Now that I know Jen has 15 bracelets left, I can split this part into 15 smaller equal parts. I know that she sells each bracelet for $3, so each part has a value of $3. I can also label the unknown as *m* to represent how much money Jen earns in total.

m **represents the amount of money Jen earns**

$$15 \times 3 = m$$
$$m = (10 \times 3) + (5 \times 3)$$
$$m = 30 + 15$$
$$m = 45$$

I can write what *m* represents and then write an equation to solve for *m*. I need to multiply 15 by 3, a large fact! I can use the break apart and distribute strategy to solve for 15 × 3. I can break up 15 threes as 10 threes and 5 threes and then find the sum of the 2 smaller facts.

Jen earns a total of $45 *from selling* 15 *bracelets.*

Jen does not have enough money to buy the art set. She is $10 *short.*

I am not finished answering the question until I explain why Jen does not have enough money to buy the art set.

Lesson 21:   Solve two-step word problems involving multiplying single-digit factors and multiplies of 10.

©2018 Great Minds®. eureka-math.org

EUREKA MATH™

Name _____ Date _____

Use the RDW process for each problem. Use a letter to represent the unknown.

1. There are 60 minutes in 1 hour. Use a tape diagram to find the total number of minutes in 6 hours and 15 minutes.

2. Ms. Lemus buys 7 boxes of snacks. Each box has 12 packets of fruit snacks and 18 packets of cashews. How many snack packets does she buy altogether?

3. Tamara wants to buy a tablet that costs $437. She saves $50 a month for 9 months. Does she have enough money to buy the tablet? Explain why or why not.

EUREKA MATH

Lesson 21: Solve two-step word problems involving multiplying single-digit factors and multiplies of 10.

259

©2018 Great Minds®. eureka-math.org

4. Mr. Ramirez receives 4 sets of books. Each set has 16 fiction books and 14 nonfiction books. He puts 97 books in his library and donates the rest. How many books does he donate?

5. Celia sells calendars for a fundraiser. Each calendar costs $9. She sells 16 calendars to her family members and 14 calendars to the people in her neighborhood. Her goal is to earn $300. Does Celia reach her goal? Explain your answer.

6. The video store sells science and history movies for $5 each. How much money does the video store make if it sells 33 science movies and 57 history movies?

Lesson 21: Solve two-step word problems involving multiplying single-digit factors and multiplies of 10.

EUREKA
MATH

Grade 3
Module 4

1.   Vivian uses squares to find the area of a rectangle. Her work is shown below.

   a.   How many squares did she use to cover the rectangle?

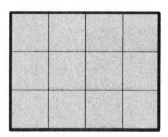

I know that the amount of flat space a shape takes up is called its area.

_____**12**_____ squares

I know these are called square units because the units used to measure area are squares. I also know that to measure area there shouldn't be any gaps or overlaps.

   b.   What is the area of the rectangle in square units? Explain how you found your answer.

   The area of the rectangle is 12 square units. I know because I counted 12 squares inside the rectangle.

2.   Each ☐ is 1 square unit. Which rectangle has the largest area? How do you know?

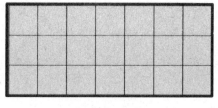

I can compare the areas of these rectangles because the same-sized square unit is used to cover each one.

Rectangle A

21 *square units*

Rectangle B

12 *square units*

Rectangle A has the largest area. I know because I counted the square units in each rectangle. Rectangle A needs the most squares to cover it with no gaps or overlaps.

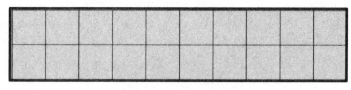

Rectangle C

20 *square units*

Name ___Tushy_____   Date _____

1. Magnus covers the same shape with triangles, rhombuses, and trapezoids.

    a. How many triangles will it take to cover the shape?

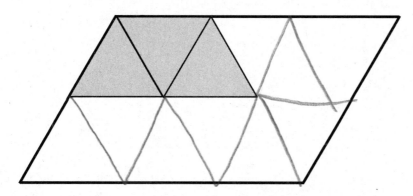

    _12_ triangles

    b. How many rhombuses will it take to cover the shape?

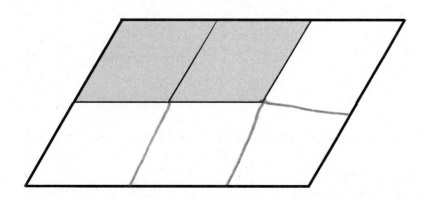

    _6_ rhombuses

    c. Magnus notices that 3 triangles from Part (a) cover 1 trapezoid. How many trapezoids will you need to cover the shape below? Explain your answer.

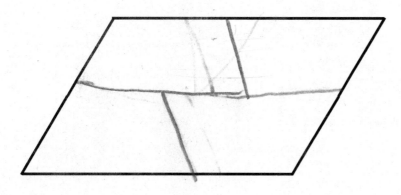

    _4_ trapezoids

2. Angela uses squares to find the area of a rectangle. Her work is shown below.

   a. How many squares did she use to cover the rectangle?

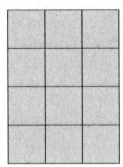

   _12_ squares

   b. What is the area of the rectangle in square units? Explain how you found your answer.

   12 sq. units
   I skip counted.

3. Each ▢ is 1 square unit. Which rectangle has the largest area? How do you know?

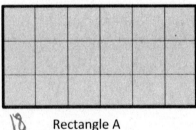

Ⓐ
I skip counted

18   Rectangle A

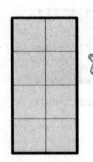

8

Rectangle B

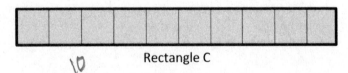

10   Rectangle C

Lesson 1:   Understand area as an attribute of plane figures.

©2018 Great Minds®. eureka-math.org

EUREKA MATH

1. Matthew uses square inches to create these rectangles. Do they have the same area? Explain.

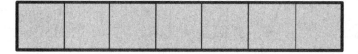

7 square inches

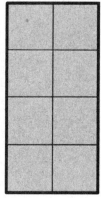

8 square inches

No, they do not have the same area. I counted the square inches in each rectangle and found that the rectangle on the right has a larger area by 1 square inch.

This is the new unit I learned today. Each side of a square inch measures 1 inch. The units in this drawing are just meant to represent square inches. I can write square inches as sq in for short.

2. Each ☐ is a square unit. Count to find the area of the rectangle below. Then, draw a different rectangle that has the same area.

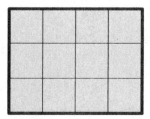

12 square units

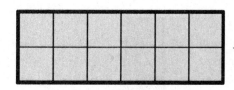

12 square units

I can rearrange the 12 square units into two equal rows to make a new rectangle. I know that rearranging the square units does not change the area because no new units are added, and none are taken away.

Name __Tvisha_____ Date _____

1. Each is a square unit. Count to find the area of each rectangle. Then, circle all the rectangles with an area of 12 square units.

a.

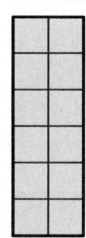

Area = __12__ square units

b.

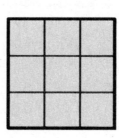

Area = __9__ square units

c.

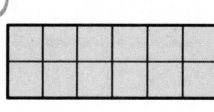

Area = __12__ square units

d.

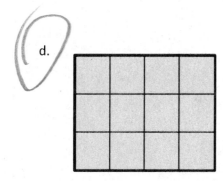

Area = __12__ square units

e.

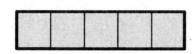

Area = __5__ square units

f.

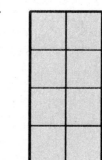

Area = __8__ square units

EUREKA MATH™

©2018 Great Minds®. eureka-math.org

2.   Colin uses square units to create these rectangles. Do they have the same area? Explain.

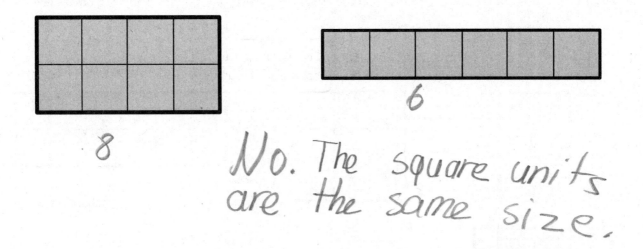

8   6

No. The square units are the same size.

3.   Each ☐ is a square unit. Count to find the area of the rectangle below. Then, draw a different rectangle that has the same area.

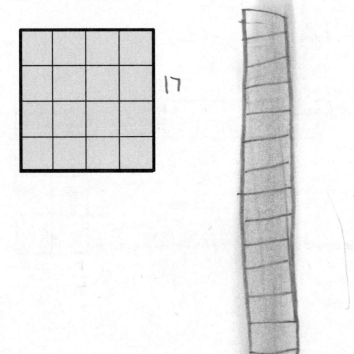

17

EUREKA
MATH™

1. Each ☐ is 1 square unit. What is the area of each of the following rectangles?

   a.

> I can find the area of each rectangle by counting the number of square units.

   6 square units

   b.

   20 square units

2. How would the rectangles in Problem 1 be different if they were composed of square inches?

The number of squares in each rectangle would stay the same, but the side of each square would measure 1 inch. We would also label the area as square inches instead of square units.

> I know 1 square inch covers a greater area than 1 square centimeter because 1 inch is longer than 1 centimeter.

3. How would the rectangles in Problem 1 be different if they were composed of square centimeters?

The number of squares in each rectangle would stay the same, but the side of each square would measure 1 centimeter. We would also label the area as square centimeters instead of square units.

Lesson 3: Model tiling with centimeter and inch unit lsquares as a strategy to measure area.

©2018 Great Minds®. eureka-math.org

271

Name __Tvisha_____   Date _____

1. Each ☐ is 1 square unit. What is the area of each of the following rectangles?

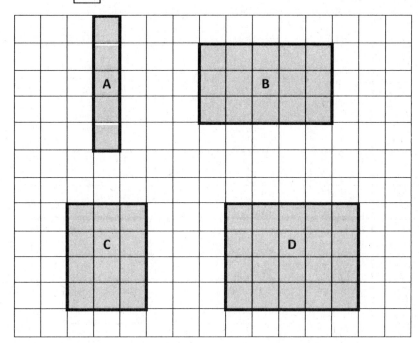

A: ___5___ square units

B: __15 sq.units__

C: __12 sq.units__

D: __20 sq. units__

2. Each ☐ is 1 square unit. What is the area of each of the following rectangles?

a.

_____9 sq. u_____

b.

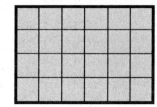

_____24 sq.u_____

c.

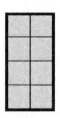

_____8 sq. u_____

d.

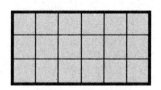

_____18 sq.u_____

EUREKA MATH™

Lesson 3: Model tiling with centimeter and inch unit squares as a strategy to measure area.

©2018 Great Minds®. eureka-math.org

273

3. Each ☐ is 1 square unit. Write the area of each rectangle. Then, draw a different rectangle with the same area in the space provided.

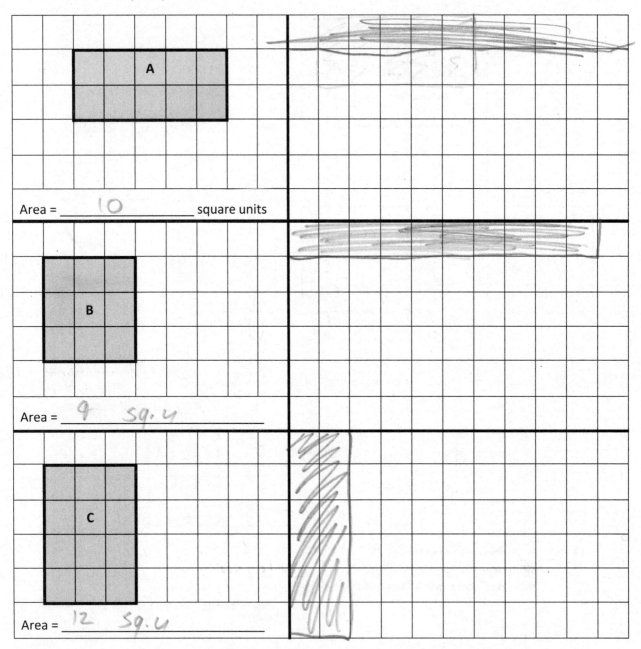

Area = _____10_____ square units

Area = __9__ sq. u_____

Area = __12__ sq. u_____

Lesson 3: Model tiling with centimeter and inch unit squares as a strategy to measure area.

©2018 Great Minds®. eureka-math.org

EUREKA MATH

1. Use a ruler to measure the side lengths of the rectangle in centimeters. Mark each centimeter with a point, and draw lines from the points to show the square units. Then, count the squares you drew to find the total area.

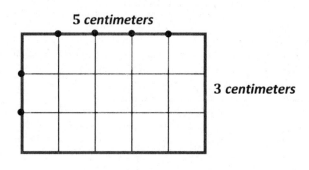

5 centimeters

3 centimeters

> I know the side length of a rectangle is the same as the number of centimeter tiles that make it. I also know that opposite sides of rectangles are equal, so I only need to measure 2 sides.

Total area: __15 square centimeters__

2. Each ☐ is 1 square centimeter. Sammy says that the side length of the rectangle below is 8 centimeters. Davis says the side length is 3 centimeters. Who is correct? Explain how you know.

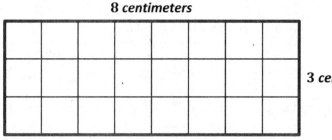

8 centimeters

3 centimeters

> An efficient strategy to find the area is to think of this rectangle as 3 rows of 8 tiles, or 3 eights. Then we can skip-count by eights 3 times to find the total number of square centimeter tiles.

They are both correct because I counted the tiles across the top, and there are 8 tiles, which means that the side length is 8 cm. Then I counted the tiles along the side, and there are 3 tiles, which means that the side length is 3 cm.

Lesson 4:   Relate side lengths with the number of tiles on a side.

©2018 Great Minds®. eureka-math.org

275

3. Shana uses square inch tiles to find the side lengths of the rectangle below. Label each side length. Then, find the total area.

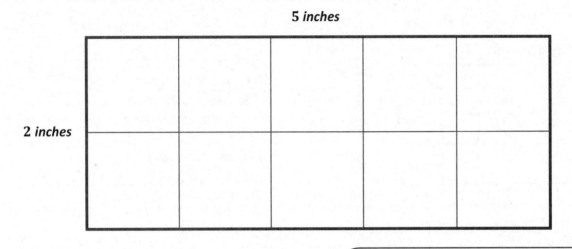

5 *inches*

2 *inches*

Total area: __10 *square inches*__

> I know the units are labeled differently for side lengths and area. I know the unit for side lengths is inches because the unit measures the length of the side in inches. For area, the unit is square inches because I count the number of square inch tiles that are used to make the rectangle.

4. How does knowing side lengths W and X help you find side lengths Y and Z on the rectangle below?

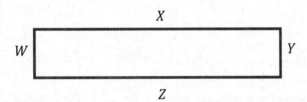

X

W Y

Z

I know that opposite sides of a rectangle are equal. So, if I know side length X, I also know side length Z. If I know side length W, I also know side length Y.

Lesson 4:   Relate side lengths with the number of tiles on a side.

©2018 Great Minds®. eureka-math.org

EUREKA
MATH

Name _____   Date _____

1. Ella placed square centimeter tiles on the rectangle below, and then labeled the side lengths. What is the area of her rectangle?

4 cm

2 cm

Total area: _____

2. Kyle uses square centimeter tiles to find the side lengths of the rectangle below. Label each side length. Then, count the tiles to find the total area.

Total area: _____

3. Maura uses square inch tiles to find the side lengths of the rectangle below. Label each side length. Then, find the total area.

Total area: _____

4. Each square unit below is 1 square inch. Claire says that the side length of the rectangle below is 3 inches. Tyler says the side length is 5 inches. Who is correct? Explain how you know.

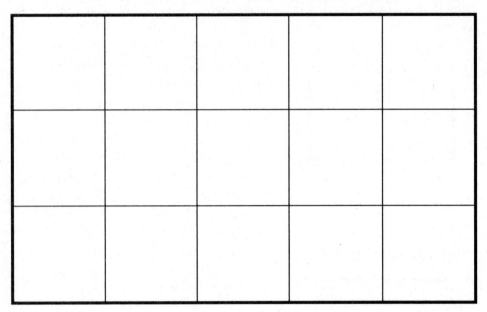

5. Label the unknown side lengths for the rectangle below, and then find the area. Explain how you used the lengths provided to find the unknown lengths and area.

4 inches

2 inches

_____

_____

Total area: _____

Lesson 4:   Relate side lengths with the number of tiles on a side.

©2018 Great Minds®. eureka-math.org

EUREKA
MATH™

1. Use the centimeter side of a ruler to draw in the tiles. Then, find and label the unknown side length. Skip-count the tiles to check your work. Write a multiplication sentence for each tiled rectangle.

a. Area: 12 square centimeters

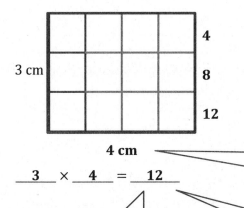

4 cm

3
6
9
12

3 cm

__4__ × __3__ = __12__

> I can use my ruler to mark each centimeter. Then, I can connect the marks to draw the tiles. I'll count the square units and label the unknown side length 3 cm.

> Next, I'll skip-count by 3 to check that the total number of tiles matches the given area of 12 square centimeters.

> I can write 3 for the unknown factor because my tiled array shows 4 rows of 3 tiles.

b. Area: 12 square centimeters

3 cm

4
8
12

4 cm

__3__ × __4__ = __12__

> After I use my ruler to draw the tiles, I can count to find the unknown side length and label it.

> I can write the number sentence 3 × 4 = 12 because there are 3 rows of 4 tiles, which is a total of 12 tiles.

> The area of the rectangles in parts (a) and (b) is 12 square centimeters. That means both rectangles have the same area even though they look different.

Lesson 5:   Form rectangles by tiling with unit squares to make arrays.

EUREKA MATH™

©2018 Great Minds®. eureka-math.org

279

2. Ella makes a rectangle with 24 square centimeter tiles. There are 4 equal rows of tiles.

a. How many tiles are in each row? Use words, pictures, and numbers to support your answer.

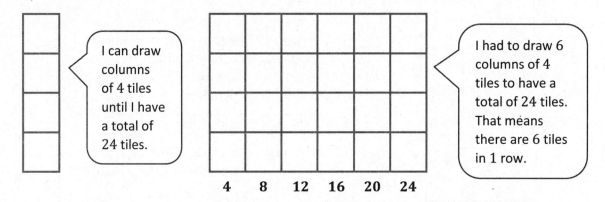

I can draw columns of 4 tiles until I have a total of 24 tiles.

I had to draw 6 columns of 4 tiles to have a total of 24 tiles. That means there are 6 tiles in 1 row.

4 8 12 16 20 24

There are 6 tiles in each row. I drew columns of 4 tiles until I had a total of 24 tiles. Then I counted how many tiles are in 1 row. I could also find the answer by thinking about the problem as 4 × _____ = 24 because I know that 4 × 6 = 24.

b. Can Ella arrange all of her 24 square centimeter tiles into 3 equal rows? Use words, pictures, and numbers to support your answer.

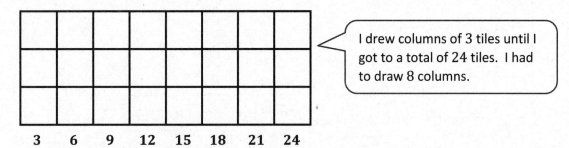

I drew columns of 3 tiles until I got to a total of 24 tiles. I had to draw 8 columns.

3 6 9 12 15 18 21 24

Yes, Ella can arrange all of her 24 tiles into 3 equal rows. I drew columns of 3 tiles until I had a total of 24 tiles. I can use my picture to see that there are 8 tiles in each row. I can also use multiplication to help me because I know that 3 × 8 = 24.

c. Do the rectangles in parts (a) and (b) have the same total area? Explain how you know.

Yes, the rectangles in parts (a) and (b) have the same area because they are both made up of 24 square centimeter tiles. The rectangles look different because they have different side lengths, but they have the same area.

This is different than Problem 1 because the rectangles in Problem 1 had the same side lengths. They were just rotated.

Lesson 5:   Form rectangles by tiling with unit squares to make arrays.

Name _____   Date _____

1. Use the centimeter side of a ruler to draw in the tiles. Find the unknown side length or skip-count to find the unknown area. Then, complete the multiplication equations.

a. Area: **24** square centimeters.

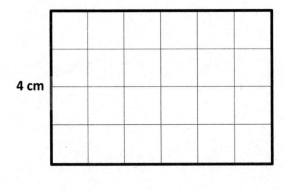

_____4_____ × _____ = ____24____

b. Area: **24** square centimeters.

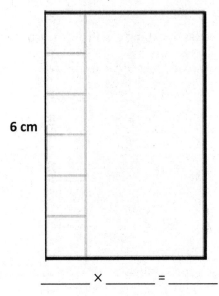

_____ × _____ = _____

c. Area: **15** square centimeters.

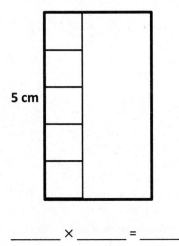

_____ × _____ = _____

d. Area: **15** square centimeters.

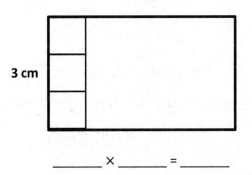

_____ × _____ = _____

Lesson 5:   Form rectangles by tiling with unit squares to make arrays.

EUREKA
MATH™

©2018 Great Minds®. eureka-math.org

281

2. Ally makes a rectangle with 45 square inch tiles. She arranges the tiles in 5 equal rows. How many square inch tiles are in each row? Use words, pictures, and numbers to support your answer.

3. Leon makes a rectangle with 36 square centimeter tiles. There are 4 equal rows of tiles.

   a. How many tiles are in each row? Use words, pictures, and numbers to support your answer.

   b. Can Leon arrange all of his 36 square centimeter tiles into 6 equal rows? Use words, pictures, and numbers to support your answer.

   c. Do the rectangles in Parts (a) and (b) have the same total area? Explain how you know.

Lesson 5:   Form rectangles by tiling with unit squares to make arrays.

1. Each ☐ represents 1 square centimeter. Draw to find the number of rows and columns in each array. Match it to its completed array. Then, fill in the blanks to make a true equation to find each array's area.

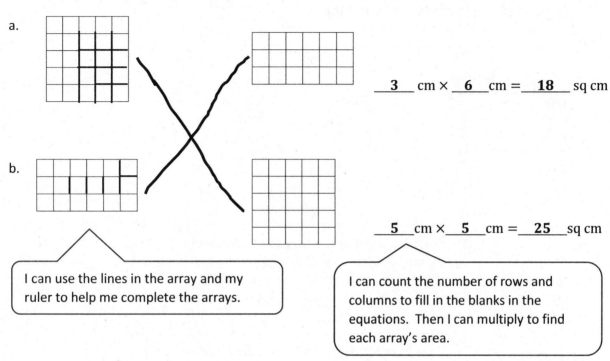

a.

    __3__ cm × __6__ cm = __18__ sq cm

b.

    __5__ cm × __5__ cm = __25__ sq cm

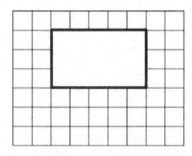

I can use the lines in the array and my ruler to help me complete the arrays.

I can count the number of rows and columns to fill in the blanks in the equations. Then I can multiply to find each array's area.

2. A painting covers the tile wall in Ava's kitchen, as shown below.

a. Ava skip-counts by 9 to find the total number of square tiles on the wall. She says there are 63 square tiles. Is she correct? Explain your answer.

Yes, Ava is correct. Even though I can't see all of the tiles, I can use the first row and column to see that there are 7 rows of 9 tiles. I can multiply 7 × 9, which equals 63.

Lesson 6:   Draw rows and columns to determine the area of a rectangle given an incomplete array.

©2018 Great Minds®. eureka-math.org

283

EUREKA
MATH™

b.   How many square tiles are under the painting?

I can use the tiles around the painting to help me figure out how many tiles are under the painting.

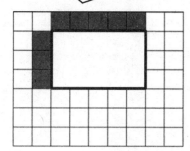

$3 \times 5 = 15$

There are 3 rows of square tiles and 5 columns of square tiles under the painting. I can multiply 3×5 to find the total number of tiles under the painting.

$63 - 48 = 15$

I know from part (a) that there are 63 total tiles. So, I could also solve by subtracting the number of tiles that I can see from the total.

There are 15 *square tiles under the painting.*

Lesson 6:   Draw rows and columns to determine the area of a rectangle given an incomplete array.

Name _____ Date _____

1. Each ☐ represents 1 square centimeter. Draw to find the number of rows and columns in each array. Match it to its completed array. Then, fill in the blanks to make a true equation to find each array's area.

a.

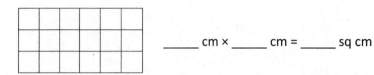

_____ cm × _____ cm = _____ sq cm

b.

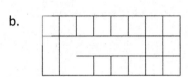

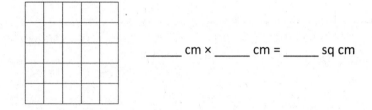

_____ cm × _____ cm = _____ sq cm

c.

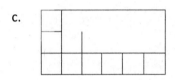

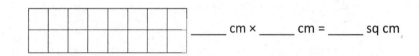

_____ cm × _____ cm = _____ sq cm

d.

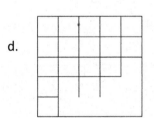

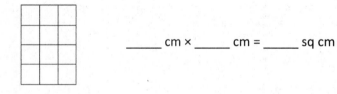

_____ cm × _____ cm = _____ sq cm

e.

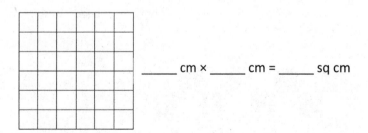

_____ cm × _____ cm = _____ sq cm

f.

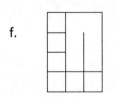

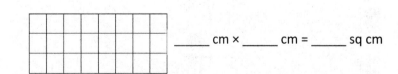

_____ cm × _____ cm = _____ sq cm

EUREKA
MATH™

Lesson 6: Draw rows and columns to determine the area of a rectangle given an incomplete array.

©2018 Great Minds®. eureka-math.org

285

2. Minh skip-counts by sixes to find the total square units in the rectangle below. She says there are 36 square units. Is she correct? Explain your answer.

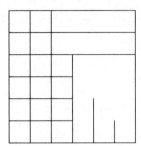

3. The tub in Paige's bathroom covers the tile floor as shown below. How many square tiles are on the floor, including the tiles under the tub?

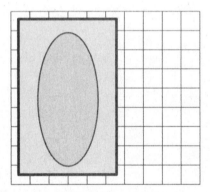

4. Frank sees a notebook on top of his chessboard. How many squares are covered by the notebook? Explain your answer.

Lesson 6: Draw rows and columns to determine the area of a rectangle given an incomplete array.

1. Find the area of the rectangular array. Label the side lengths of the matching area model, and write a multiplication equation for the area model.

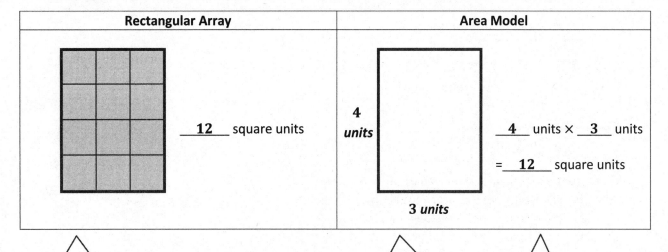

Rectangular Array	Area Model
___**12**___ square units	**4 units** **3 units** ___**4**___ units × ___**3**___ units =___**12**___ square units

I can skip-count rows by 3 or columns by 4 to find the area of the rectangular array.

I can use the rectangular array to help me label the side lengths of the area model. There are 4 rows, so the width is 4 units. There are 3 columns, so the length is 3 units.

I can multiply 4 × 3 to find the area. The area model and the rectangular array have the same area of 12 square units.

2. Mason arranges square pattern blocks into a 3 by 6 array. Draw Mason's array on the the grid below. How many square units are in Mason's rectangular array?

a.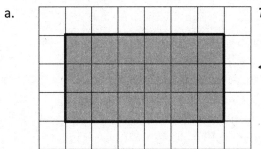

There are 18 square units in Mason's rectangular array.

I can draw a rectangular array with 3 rows and 6 columns. Then I can multiply 3 × 6 to find the total number of square units in the rectangular array.

b. Label the side lengths of Mason's array from part (a) on the rectangle below. Then, write a multiplication sentence to represent the area of the rectangle.

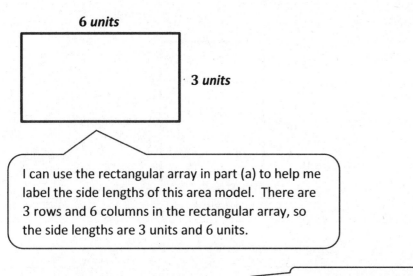

6 units

3 units

I can use the rectangular array in part (a) to help me label the side lengths of this area model. There are 3 rows and 6 columns in the rectangular array, so the side lengths are 3 units and 6 units.

3 units × 6 units = 18 square units

I can multiply the side lengths to find the area.

3. Luke draws a rectangle that is 4 square feet. Savannah draws a rectangle that is 4 square inches. Whose rectangle is larger in area? How do you know?

Luke's rectangle is larger in area because they both used the same number of units, but the size of the units is different. Luke used square feet, which are larger than square inches. Since the units that Luke used are larger than the units that Savannah used and they both used the same number of units, Luke's rectangle is larger in area.

I can think about the lesson today to help me answer this question. My partner and I made rectangles using square inch and square centimeter tiles. We both used the same number of tiles to make our rectangles, but we noticed that the rectangle made of square inches was larger in area than the rectangle made of square centimeters. The larger unit, square inches, made a rectangle with a larger area.

Lesson 7:   Interpret area models to form rectangular arrays.

©2018 Great Minds®. eureka-math.org

EUREKA MATH

Name _____ Date _____

1. Find the area of each rectangular array. Label the side lengths of the matching area model, and write a multiplication equation for each area model.

Rectangular Arrays	Area Models
a. _____ square units	3 units 3 units × _____ units = _____ square units 2 units
b. _____ square units	_____ units × _____ units = _____ square units
c. _____ square units	_____ units × _____ units = _____ square units
d. _____ square units	_____ units × _____ units = _____ square units

2. Jillian arranges square pattern blocks into a 7 by 4 array. Draw Jillian's array on the the grid below. How many square units are in Jillian's rectangular array?

a.

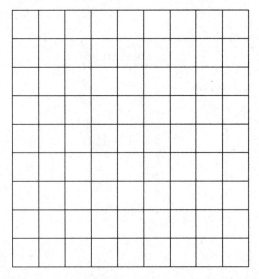

b. Label the side lengths of Jillian's array from Part (a) on the rectangle below. Then, write a multiplication sentence to represent the area of the rectangle.

3. Fiona draws a 24 square centimeter rectangle. Gregory draws a 24 square inch rectangle. Whose rectangle is larger in area? How do you know?

Lesson 7: Interpret area models to form rectangular arrays.

EUREKA
MATH

1. Write a multiplication equation to find the area of the rectangle.

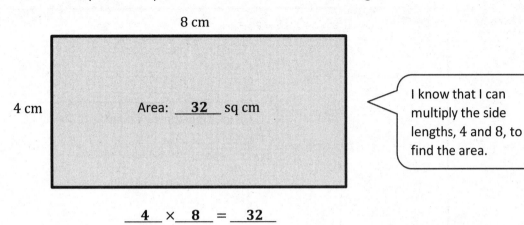

8 cm

4 cm   Area: __32__ sq cm

> I know that I can multiply the side lengths, 4 and 8, to find the area.

__4__ × __8__ = __32__

2. Write a multiplication equation and a division equation to find the unknown side length for the rectangle.

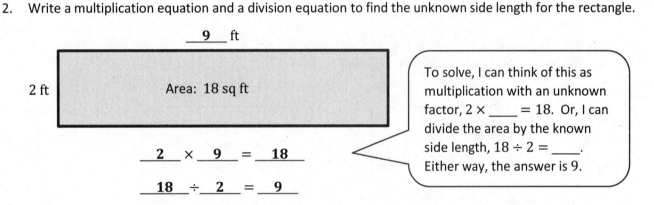

__9__ ft

2 ft   Area: 18 sq ft

__2__ × __9__ = __18__

__18__ ÷ __2__ = __9__

> To solve, I can think of this as multiplication with an unknown factor, 2 × ____ = 18. Or, I can divide the area by the known side length, 18 ÷ 2 = ____. Either way, the answer is 9.

3. On the grid below, draw a rectangle that has an area of 24 square units. Label the side lengths.

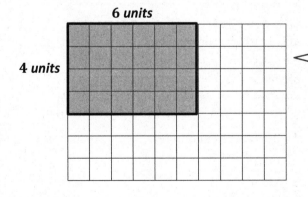

6 units

4 units

> To draw a rectangle with an area of 24 square units, I can think about factors of 24. I know 4 × 6 = 24, so my side lengths can be 4 and 6.

Lesson 8:   Find the area of a rectangle through multiplication of the side lengths.

EUREKA MATH™

©2018 Great Minds®. eureka-math.org

291

4. Keith draws a rectangle that has side lengths of 6 inches and 3 inches. What is the area of the rectangle? Explain how you found your answer.

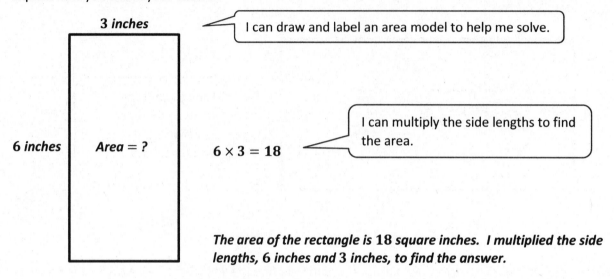

I can draw and label an area model to help me solve.

I can multiply the side lengths to find the area.

$6 \times 3 = 18$

The area of the rectangle is 18 square inches. I multiplied the side lengths, 6 inches and 3 inches, to find the answer.

5. Isabelle draws a rectangle with a side length of 5 centimeters and an area of 30 square centimeters. What is the other side length? How do you know?

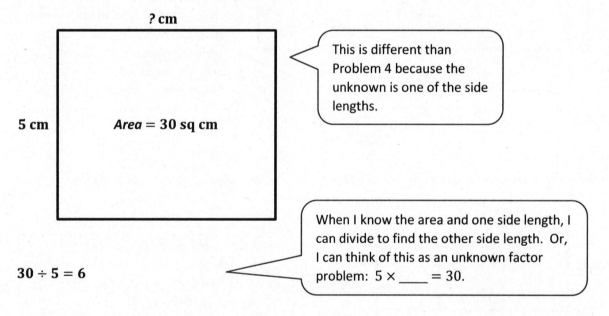

This is different than Problem 4 because the unknown is one of the side lengths.

When I know the area and one side length, I can divide to find the other side length. Or, I can think of this as an unknown factor problem: $5 \times$ ____ $= 30$.

$30 \div 5 = 6$

The other side length is 6 centimeters. I divided the area, 30 square centimeters, by the known side length, 5 centimeters, and $30 \div 5 = 6$.

Lesson 8:   Find the area of a rectangle through multiplication of the side lengths.

EUREKA MATH

Name _____   Date _____

1. Write a multiplication equation to find the area of each rectangle.

a.

8 cm

3 cm | Area: _____ sq cm

_____ × _____ = _____

b.

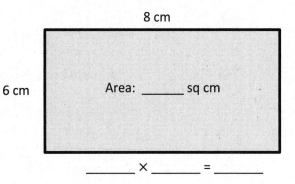

8 cm

6 cm   Area: _____ sq cm

_____ × _____ = _____

c.

4 ft

4 ft | Area: _____ sq ft

_____ × _____ = _____

d.

7 ft

4 ft   Area: _____ sq ft

_____ × _____ = _____

2. Write a multiplication equation and a division equation to find the unknown side length for each rectangle.

a.

_____ ft.

3 ft | Area: 24 sq ft

_____ × _____ = _____

_____ ÷ _____ = _____

b.

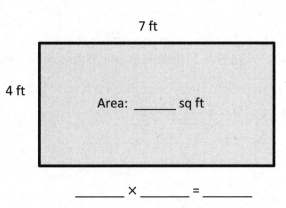

9 ft

_____ ft   Area: 36 sq ft

_____ × _____ = _____

_____ ÷ _____ = _____

Lesson 8:   Find the area of a rectangle through multiplication of the side lengths.

EUREKA
MATH™

©2018 Great Minds®. eureka-math.org

293

3. On the grid below, draw a rectangle that has an area of 32 square centimeters. Label the side lengths.

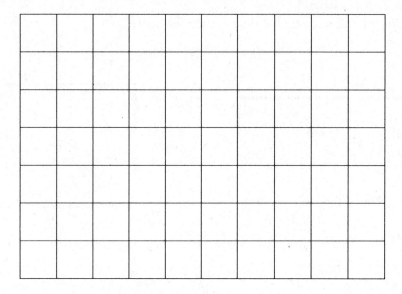

4. Patricia draws a rectangle that has side lengths of 4 centimeters and 9 centimeters. What is the area of the rectangle? Explain how you found your answer.

5. Charles draws a rectangle with a side length of 9 inches and an area of 27 square inches. What is the other side length? How do you know?

Lesson 8:   Find the area of a rectangle through multiplication of the side lengths.

EUREKA
MATH™

1. Use the grid to answer the questions below.

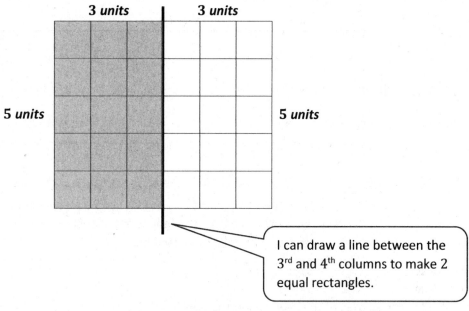

> I can draw a line between the 3rd and 4th columns to make 2 equal rectangles.

a. Draw a line to divide the grid into 2 equal rectangles. Shade in 1 of the rectangles that you created.

> I can count the units on each side to help me label the side lengths of each rectangle.

b. Label the side lengths of each rectangle.

c. Write an equation to show the total area of the 2 rectangles.

Area $= (5 \times 3) + (5 \times 3)$
$= 15 + 15$
$= 30$

The total area is 30 square units.

> I can find the area of each smaller rectangle by multiplying 5×3. Then, I can add the areas of the 2 equal rectangles to find the total area.

Lesson 9:   Analyze different rectangles and reason about their area.

EUREKA MATH™

©2018 Great Minds®. eureka-math.org

295

2. Phoebe cuts out the 2 equal rectangles from Problem 1(a) and puts the two shorter sides together.

   a. Draw Phoebe's new rectangle, and label the side lengths below.

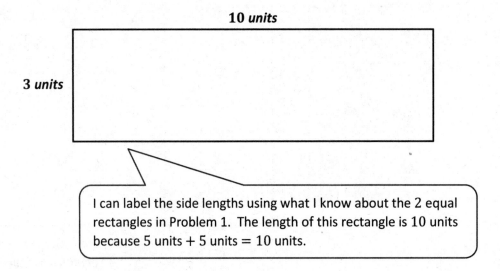

10 units

3 units

I can label the side lengths using what I know about the 2 equal rectangles in Problem 1. The length of this rectangle is 10 units because 5 units + 5 units = 10 units.

   b. Find the total area of the new, longer rectangle.

   $Area = 3 \times 10$
   $= 30$

   The total area is 30 square units.

   I can find the area by multiplying the side lengths.

   c. Is the area of the new, longer rectangle equal to the total area in Problem 1(c)? Explain why or why not.

   Yes, the area of the new, longer rectangle is equal to the total area in Problem 1(c). Phoebe just rearranged the 2 smaller, equal rectangles, so the total area didn't change.

   I know that the total area doesn't change just because the 2 equal rectangles were moved around to form a new, longer rectangle. No units were taken away and none were added, so the area stays the same.

Lesson 9:   Analyze different rectangles and reason about their area.

296

©2018 Great Minds®. eureka-math.org

EUREKA MATH™

Name _____   Date _____

1. Use the grid to answer the questions below.

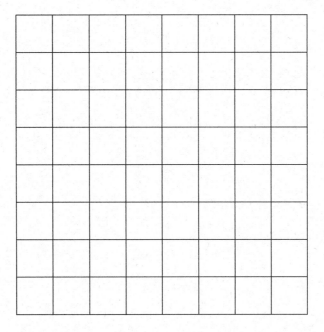

a. Draw a line to divide the grid into 2 equal rectangles. Shade in 1 of the rectangles that you created.

b. Label the side lengths of each rectangle.

c. Write an equation to show the total area of the 2 rectangles.

Lesson 9:   Analyze different rectangles and reason about their area.

297

2. Alexa cuts out the 2 equal rectangles from Problem 1(a) and puts the two shorter sides together.

   a. Draw Alexa's new rectangle and label the side lengths below.

   b. Find the total area of the new, longer rectangle.

   c. Is the area of the new, longer rectangle equal to the total area in Problem 1(c)?
      Explain why or why not.

Lesson 9:   Analyze different rectangles and reason about their area.

1. Label the side lengths of the shaded and unshaded rectangles. Then, find the total area of the large rectangle by adding the areas of the 2 smaller rectangles.

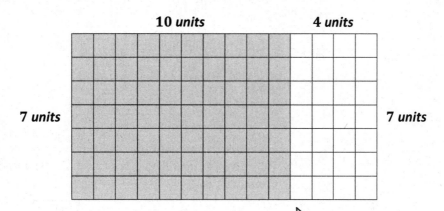

$7 \times 14 = 7 \times ($ __10__ $+$ __4__ $)$

$\qquad = (7 \times$ __10__ $) + (7 \times$ __4__ $)$

$\qquad =$ __70__ $+$ __28__

$\qquad =$ __98__

Area: __98__ square units

I can count the units on each side to help me label the side lengths of each rectangle.

Lesson 10:   Apply the distributive property as a strategy to find the total area of a large rectangle by adding two products.

©2018 Great Minds®. eureka-math.org

299

EUREKA
MATH™

2. Vickie imagines 1 more row of seven to find the total area of a 9 × 7 rectangle. Explain how this could help her solve 9 × 7.

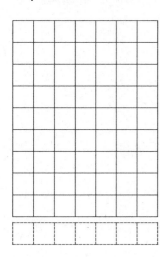

This can help her solve 9 × 7 because now she can think of it as 10 × 7 minus 1 seven. 10 × 7 might be easier for Vickie to solve than 9 × 7.

10 × 7 = 70

70 − 7 = 63

This reminds me of the 9 = 10 − 1 strategy that I can use to multiply by 9.

3. Break the 16 × 6 rectangle into 2 rectangles by shading one smaller rectangle within it. Then, find the total area by finding the sum of the areas of the 2 smaller rectangles. Explain your thinking.

6 units

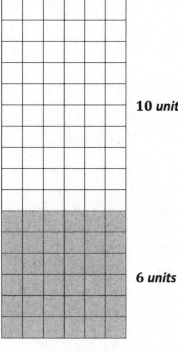

10 units

6 units

Area = $(10 × 6) + (6 × 6)$

      = $60 + 36$

      = 96

The total area is 96 square units.

I broke apart the 16 × 6 rectangle into 2 smaller rectangles: 10 × 6 and 6 × 6. I chose to break it apart like this because those are easy facts for me. I multiplied the side lengths to find the area of each smaller rectangle and added those areas to find the total area.

I can break apart the rectangle any way I want to, but I like to look for facts that are easy for me to solve. Multiplying by 10 is easy for me. I also could have broken it apart into 8 × 6 and 8 × 6. Then I would really only have to solve one fact.

Lesson 10:   Apply the distributive property as a strategy to find the total area of a large rectangle by adding two products.

©2018 Great Minds®. eureka-math.org

EUREKA MATH

Name _____ Date _____

1. Label the side lengths of the shaded and unshaded rectangles. Then, find the total area of the large rectangle by adding the areas of the 2 smaller rectangles.

a.

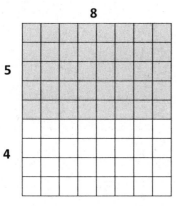

$9 \times 8 = (5 + 4) \times 8$

$= (5 \times 8) + (4 \times 8)$

$= \rule{2cm}{0.4pt} + \rule{2cm}{0.4pt}$

$= \rule{2cm}{0.4pt}$

Area: _____ square units

b.

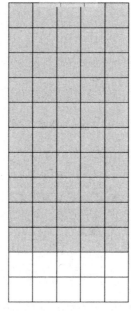

$12 \times 5 = (\rule{1.5cm}{0.4pt} + 2) \times 5$

$= (\rule{1.5cm}{0.4pt} \times 5) + (2 \times 5)$

$= \rule{1.5cm}{0.4pt} + 10$

$= \rule{1.5cm}{0.4pt}$

Area: _____ square units

c.

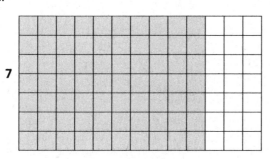

$7 \times 13 = 7 \times (\rule{1.5cm}{0.4pt} + 3)$

$= (7 \times \rule{1.5cm}{0.4pt}) + (7 \times 3)$

$= \rule{1.5cm}{0.4pt} + \rule{1.5cm}{0.4pt}$

$= \rule{1.5cm}{0.4pt}$

Area: _____ square units

d.

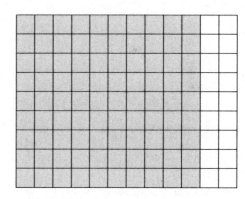

$9 \times 12 = 9 \times (\rule{1.5cm}{0.4pt} + \rule{1.5cm}{0.4pt})$

$= (9 \times \rule{1.5cm}{0.4pt}) + (9 \times \rule{1.5cm}{0.4pt})$

$= \rule{1.5cm}{0.4pt} + \rule{1.5cm}{0.4pt}$

$= \rule{1.5cm}{0.4pt}$

Area: _____ square units

Lesson 10: Apply the distributive property as a strategy to find the total area of a large rectangle by adding two products.

©2018 Great Minds®. eureka-math.org

301

2. Finn imagines 1 more row of nine to find the total area of 9 × 9 rectangle. Explain how this could help him solve 9 × 9.

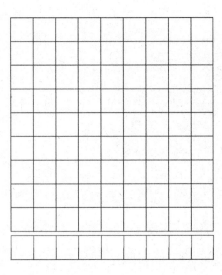

3. Shade an area to break the 16 × 4 rectangle into 2 smaller rectangles. Then, find the sum of the areas of the 2 smaller rectangles to find the total area. Explain your thinking.

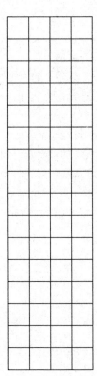

Lesson 10:   Apply the distributive property as a strategy to find the total area of a large rectangle by adding two products.

©2018 Great Minds®. eureka-math.org

EUREKA
MATH™

1. The rectangles below have the same area. Move the parentheses to find the unknown side lengths.
    Then, solve.

    a.

    6 cm

    4 cm

    Area: $4 \times$ __**6**__ $=$ __**24**__

    Area: __**24**__ sq cm

    > I can
    > multiply
    > the side
    > lengths to
    > find the
    > area.

    b.

    __**12**__ cm

    __**2**__ cm

    Area: $4 \times 6 = (2 \times 2) \times 6$
    $= 2 \times (2 \times 6)$
    $=$ __**2**__ $\times$ __**12**__
    $=$ __**24**__

    Area: __**24**__ sq cm

    > I can move the parentheses to be
    > around 2×6. After I multiply 2×6,
    > I have new side lengths of 2 cm and
    > 12 cm. I can label the side lengths
    > on the rectangle. The area didn't
    > change; it's still 24 sq cm.

2. Does Problem 1 show all the possible whole number side lengths for a rectangle with an area of 24
    square centimeters? How do you know?

    *No, Problem 1 doesn't show all possible whole number side lengths. I check by trying to multiply each
    number 1 through 10 by another number to equal 24. If I can find numbers that make 24 when I
    multiply them, then I know those are possible side lengths.*

    *I know $1 \times 24 = 24$. So 1 cm and 24 cm are possible side lengths. I already have a multiplication
    fact for 2, 2×12. I know $3 \times 8 = 24$, which means $8 \times 3 = 24$. I already have a multiplication fact
    for 4, 4×6. That also means that I have a fact for 6, $6 \times 4 = 24$. I know there's not a whole number
    that can be multiplied by 5, 7, 9, or 10 that equals 24. So besides the side lengths from Problem 1,
    other ones could be 1 cm and 24 cm or 8 cm and 3 cm.*

    > I know that I can't have side lengths that are both two-digit numbers because
    > when I multiply 2 two-digit numbers, the product is much larger than 24.

EUREKA MATH

Lesson 11:   Demonstrate the possible whole number side lengths of rectangles
with areas of 24, 36, 48, or 72 square units using the associative
property.
©2018 Great Minds®. eureka-math.org

303

3.

    a. Find the area of the rectangle below.

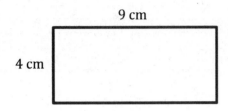

                                                    Area $= 4 \times 9$
                                                             $= 36$
                            The area of the rectangle is 36 square centimeters.

    b. Marcus says a 2 cm by 18 cm rectangle has the same area as the rectangle in part (a). Place parentheses in the equation to find the related fact and solve. Is Marcus correct? Why or why not?

$2 \times 18 = 2 \times (2 \times 9)$
          $= (2 \times 2) \times 9$
          $=$ __4__ $\times$ __9__
          $=$ __36__

Area: __36__ sq cm

Yes, Marcus is correct because I can rewrite 18 as 2×9. Then I can move the parentheses so they are around 2×2. After I multiply 2×2, I have 4 cm and 9 cm as side lengths, just like in part (a).

$2 \times 18 = 4 \times 9 = 36$

> Even though the rectangles in parts (a) and (b) have different side lengths, the areas are the same. Rewriting 18 as 2×9 and moving the parentheses helps me to see that $2 \times 18 = 4 \times 9$.

    c. Use the expression 4×9 to find different side lengths for a rectangle that has the same area as the rectangle in part (a). Show your equations using parentheses. Then, estimate to draw the rectangle and label the side lengths.

$4 \times 9 = 4 \times (3 \times 3)$
          $= (4 \times 3) \times 3$
          $= 12 \times 3$
          $= 36$

Area: **36 sq cm**

> I can rewrite 9 as 3×3. Then I can move the parentheses and multiply to find the new side lengths, 12 cm and 3 cm. I can estimate to draw the new rectangle. If I need to, I can use repeated addition, $12 + 12 + 12$, to double check that $12 \times 3 = 36$.

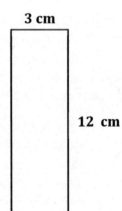

        Lesson 11:   Demonstrate the possible whole number side lengths of rectangles with areas of 24, 36, 48, or 72 square units using the associative property.
                ©2018 Great Minds®. eureka-math.org

Name _____   Date _____

1. The rectangles below have the same area. Move the parentheses to find the unknown side lengths. Then, solve.

36 cm

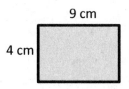

b. Area: 1 × 36 = _____

    Area: _____ sq cm

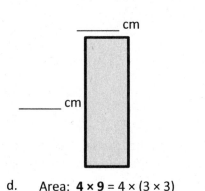

a.   Area: 4 × _____ = _____

       Area: _____ sq cm

_____ cm

2 cm

c.   Area: **4 × 9** = (2 × 2) × 9

             = 2 × 2 × 9

             = _____ × _____

             = _____

       Area: _____ sq cm

_____ cm

_____ cm

d.   Area: **4 × 9** = 4 × (3 × 3)

              = 4 × 3 × 3

              = _____ × _____

              = _____

       Area: _____ sq cm

e. Area: **12 × 3** = (6 × 2) × 3

             = 6 × 2 × 3

_____ cm

             = _____ × _____

_____ cm

             = _____

       Area: _____ sq cm

2. Does Problem 1 show all the possible whole number side lengths for a rectangle with an area of 36 square centimeters? How do you know?

Lesson 11:   Demonstrate the possible whole number side lengths of rectangles with areas of 24, 36, 48, or 72 square units using the associative property.

©2018 Great Minds®. eureka-math.org

305

EUREKA MATH™

3.   a.   Find the area of the rectangle below.

6 cm

8 cm

b.   Hilda says a 4 cm by 12 cm rectangle has the same area as the rectangle in Part (a). Place parentheses in the equation to find the related fact and solve. Is Hilda correct? Why or why not?

$$4 \times 12 = 4 \times 2 \times 6$$

$$= 4 \times 2 \times 6$$

$$= \underline{\hspace{2cm}} \times \underline{\hspace{2cm}}$$

$$= \underline{\hspace{2cm}}$$

Area: \underline{\hspace{2cm}} sq cm

c.   Use the expression 8 × 6 to find different side lengths for a rectangle that has the same area as the rectangle in Part (a). Show your equations using parentheses. Then, estimate to draw the rectangle and label the side lengths.

Lesson 11:   Demonstrate the possible whole number side lengths of rectangles with areas of 24, 36, 48, or 72 square units using the associative property.

©2018 Great Minds®. eureka-math.org

1. Molly draws a square with sides that are 8 inches long. What is the area of the square?

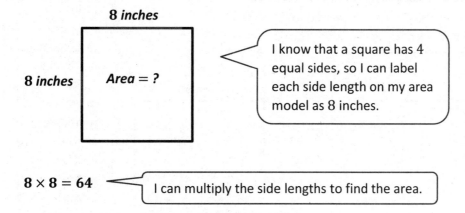

8 inches

8 inches Area = ?

> I know that a square has 4 equal sides, so I can label each side length on my area model as 8 inches.

$8 \times 8 = 64$

> I can multiply the side lengths to find the area.

The area of the square is 64 square inches.

2. Each ⬜ is 1 square unit. Nathan uses the same square units to draw a 2×8 rectangle and says that it has the same area as the rectangle below. Is he correct? Explain why or why not.

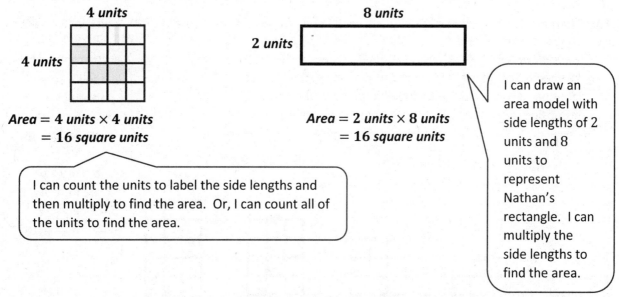

4 units

4 units

Area = 4 units × 4 units
= 16 square units

8 units

2 units

Area = 2 units × 8 units
= 16 square units

> I can count the units to label the side lengths and then multiply to find the area. Or, I can count all of the units to find the area.

> I can draw an area model with side lengths of 2 units and 8 units to represent Nathan's rectangle. I can multiply the side lengths to find the area.

Yes, Nathan is correct. Both rectangles have the same area, 16 square units. The rectangles have different side lengths, but when you multiply the side lengths, you get the same area.

$$4 \times 4 = 2 \times 8 = 16$$

EUREKA MATH™

3. A rectangular notepad has a total area of 24 square inches. Draw and label two possible notepads with different side lengths, each having an area of 24 square inches.

1×24
2×12
3×8
4×6

> I can list multiplication facts that equal 24 to help me think of possible side lengths.

> I can choose 2 facts to use as side lengths for my rectangles. I know the unit is inches because the area is in square inches.

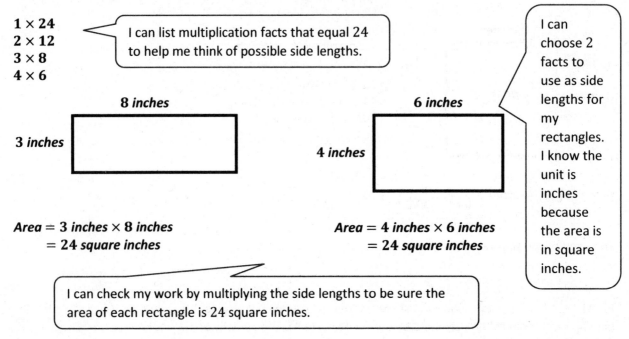

8 inches

3 inches

6 inches

4 inches

Area = **3 inches** × **8 inches**
      = **24 square inches**

Area = **4 inches** × **6 inches**
      = **24 square inches**

> I can check my work by multiplying the side lengths to be sure the area of each rectangle is 24 square inches.

4. Sophia makes the pattern below. Find and explain her pattern. Then, draw the fifth figure in her pattern.

> I can see that the first figure has 1 row of three, the second figure has 2 rows of three, and the third figure has 3 rows of three. Sophia adds 1 row of three to each new figure.

> I'll follow the pattern by drawing 4 rows of three for the fourth figure and 5 rows of three for the fifth figure.

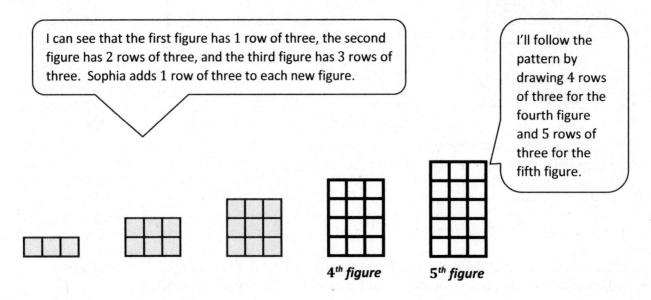

4ᵗʰ figure

5ᵗʰ figure

Sophia adds 1 row of three to each figure. The fifth figure has 5 rows of three.

Lesson 12:   Solve word problems involving area.

EUREKA MATH

Name _____   Date _____

1. A square calendar has sides that are 9 inches long. What is the calendar's area?

2. Each ☐ is 1 square unit. Sienna uses the same square units to draw a 6 × 2 rectangle and says that it has the same area as the rectangle below. Is she correct? Explain why or why not.

3. The surface of an office desk has an area of 15 square feet. Its length is 5 feet. How wide is the office desk?

4.   A rectangular garden has a total area of 48 square yards. Draw and label two possible rectangular gardens with different side lengths that have the same area.

5.   Lila makes the pattern below. Find and explain her pattern. Then, draw the *fifth* figure in her pattern.

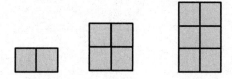

Lesson 12:   Solve word problems involving area.

©2018 Great Minds®. eureka-math.org

1. The shaded figure below is made up of 2 rectangles. Find the total area of the shaded figure.

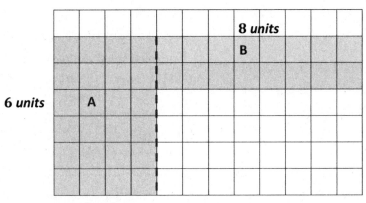

6 units

8 units

B

2 units

A

4 units

I can count the square units and label the side lengths of each rectangle inside the figure.

$6 \times 4 = 24$ $2 \times 8 = 16$

Area of A: 24 sq units   **Area of B: 16 sq units**

I can multiply the side lengths to find the area of each rectangle inside the figure.

I can add the areas of the rectangles to find the total area of the figure.

Area of A + Area of B = ___**24**___ sq units + ___**16**___ sq units = ___**40**___ sq units

6   10

$24 + 6 = 30$

$30 + 10 = 40$

I can use a number bond to help me make a ten to add. I can decompose 16 into 6 and 10. $24 + 6 = 30$ and $30 + 10 = 40$. The area of the figure is 40 square units.

Lesson 13:   Find areas by decomposing into rectangles or completing composite figures to form rectangles.

©2018 Great Minds®. eureka-math.org

311

EUREKA
MATH™

2. The figure shows a small rectangle cut out of a big rectangle. Find the area of the shaded figure.

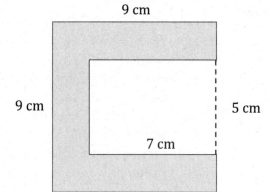

$9 \times 9 = 81$
$5 \times 7 = 35$

> I can multiply the side lengths to find the areas of the large rectangle and the unshaded rectangle.

Area of the shaded figure: __81__ − __35__ = __46__

Area of the shaded figure: __46__ square centimeters

> I can subtract the area of the unshaded rectangle from the area of the large rectangle. That helps me find just the area of the shaded figure.

3. The figure shows a small rectangle cut out of a big rectangle.

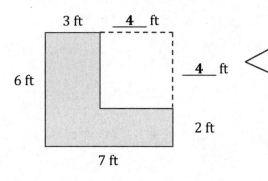

> I can label this as 4 ft because the opposite side of the rectangle is 6 ft. Since opposite sides of rectangles are equal, I can subtract the known part of this side length, 2 ft, from the opposite side length, 6 ft. 6 ft − 2 ft = 4 ft. I can use a similar strategy to find the other unknown measurement: 7 ft − 3 ft = 4 ft.

a. Label the unknown measurements.

b. Area of the big rectangle: __6__ ft × __7__ ft = __42__ sq ft

c. Area of the small rectangle: __4__ ft × __4__ ft = __16__ sq ft

d. Find the area of just the shaded part.

42 sq ft − 16 sq ft = 26 sq ft

The area of the shaded figure is 26 sq ft

> I can subtract the area of the small rectangle from the area of the big rectangle to find the area of just the shaded part.

Lesson 13: Find areas by decomposing into rectangles or completing composite figures to form rectangles.

©2018 Great Minds®. eureka-math.org

Name _____ Date _____

1. Each of the following figures is made up of 2 rectangles. Find the total area of each figure.

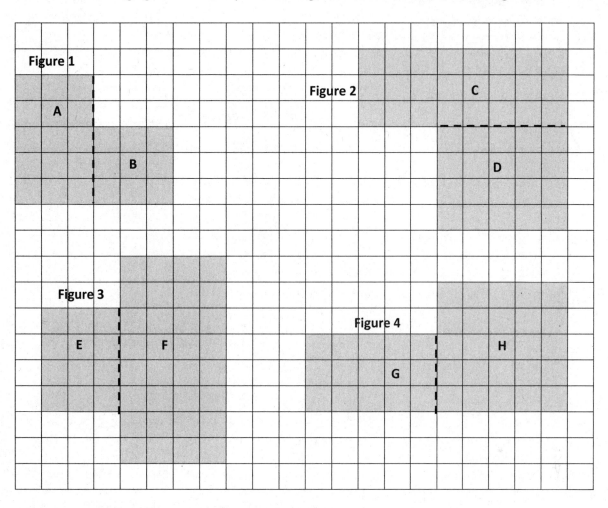

Figure 1: Area of A + Area of B: _____ sq units + _____ sq units = _____ sq units

Figure 2: Area of C + Area of D: _____ sq units + _____ sq units = _____ sq units

Figure 3: Area of E + Area of F: _____ sq units + _____ sq units = _____ sq units

Figure 4: Area of G + Area of H: _____ sq units + _____ sq units = _____ sq units

Lesson 13: Find areas by decomposing into rectangles or completing composite figures to form rectangles.

©2018 Great Minds®. eureka-math.org

313

2.   The figure shows a small rectangle cut out of a big rectangle. Find the area of the shaded figure.

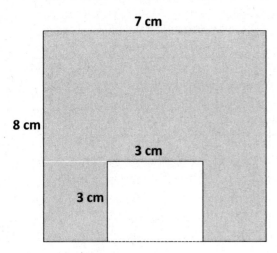

7 cm

8 cm

3 cm

3 cm

Area of the shaded figure: _____ – _____ = _____

Area of the shaded figure: _____ square centimeters

3.   The figure shows a small rectangle cut out of a big rectangle.

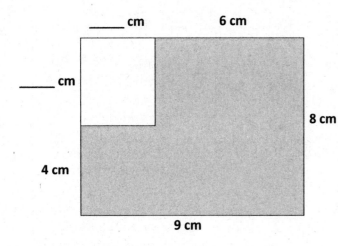

_____ cm   6 cm

_____ cm

4 cm

9 cm

8 cm

a. Label the unknown measurements.

b. Area of the big rectangle:

_____ cm × _____ cm = _____ sq cm

c. Area of the small rectangle:

_____ cm × _____ cm = _____ sq cm

d. Find the area of the shaded figure.

Lesson 13:   Find areas by decomposing into rectangles or completing composite figures to form rectangles.

©2018 Great Minds®. eureka-math.org

EUREKA
MATH™

1. Find the area of the following figure, which is made up of rectangles.

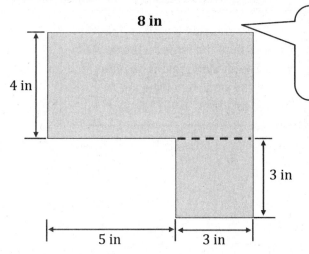

8 in

4 in

3 in

5 in   3 in

I can label this unknown side length as 8 inches because the opposite side is 5 inches and 3 inches, which makes a total of 8 inches. Opposite sides of a rectangle are equal.

$4 \times 8 = 32$

$3 \times 3 = 9$

$32 + 9 = ?$

$31 \quad 1$

$1 + 9 = 10$

$31 + 10 = 41$

I can find the area of the figure by finding the areas of the two rectangles and then adding. I can use a number bond to make adding easier.

The area of the figure is 41 square inches.

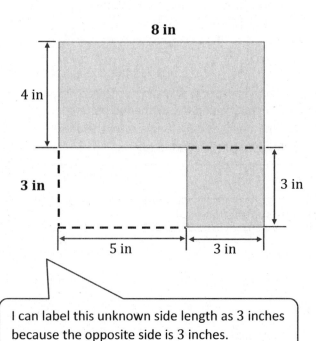

8 in

4 in

3 in

3 in

5 in   3 in

I can label this unknown side length as 3 inches because the opposite side is 3 inches.

$8 \times 7 = 56$

$3 \times 5 = 15$

$56 - 15 = 41$

Or, I can find the area of the figure by drawing lines to complete the large rectangle. Then I can find the areas of the large rectangle and the unshaded part. I can subtract the area of the unshaded part from the area of the large rectangle. Either way I solve, the area of the figure is 41 square inches.

EUREKA MATH

Lesson 14: Find areas by decomposing into rectangles or completing composite figures to form rectangles.

©2018 Great Minds®. eureka-math.org

315

2. The figure below shows a small rectangle cut out of a big rectangle. Find the area of the shaded region.

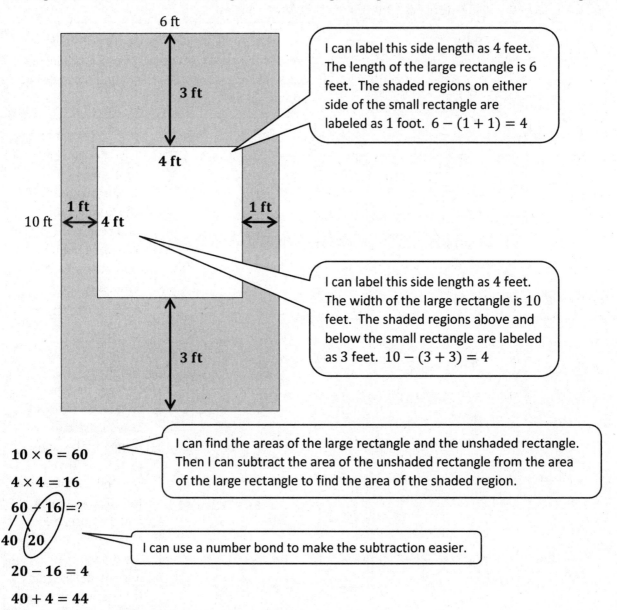

I can label this side length as 4 feet. The length of the large rectangle is 6 feet. The shaded regions on either side of the small rectangle are labeled as 1 foot. $6 - (1 + 1) = 4$

I can label this side length as 4 feet. The width of the large rectangle is 10 feet. The shaded regions above and below the small rectangle are labeled as 3 feet. $10 - (3 + 3) = 4$

$10 \times 6 = 60$

$4 \times 4 = 16$

$60 - 16 = ?$

40 20

I can find the areas of the large rectangle and the unshaded rectangle. Then I can subtract the area of the unshaded rectangle from the area of the large rectangle to find the area of the shaded region.

I can use a number bond to make the subtraction easier.

$20 - 16 = 4$

$40 + 4 = 44$

The area of the shaded region is 44 square feet.

Lesson 14: Find areas by decomposing into rectangles or completing composite figures to form rectangles.

©2018 Great Minds®. eureka-math.org

Name _____   Date _____

1. Find the area of each of the following figures. All figures are made up of rectangles.

a.

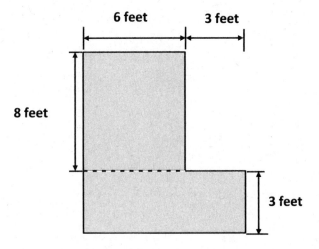

b.

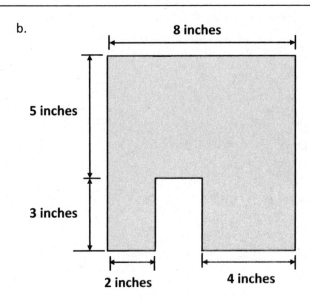

Lesson 14:   Find areas by decomposing into rectangles or completing composite
figures to form rectangles.

©2018 Great Minds®. eureka-math.org

317

2. The figure below shows a small rectangle cut out of a big rectangle.

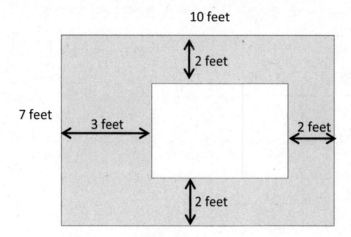

a. Label the side lengths of the unshaded region.

b. Find the area of the shaded region.

Lesson 14:   Find areas by decomposing into rectangles or completing composite
            figures to form rectangles.

©2018 Great Minds®. eureka-math.org

EUREKA
MATH™

Use a ruler to measure the side lengths of each numbered room in the floor plan in centimeters. Then, find each area. Use the measurements below to match and label the rooms.

Kitchen/Living Room: 78 square centimeters   Bedroom: 48 square centimeters

Bathroom: 24 square centimeters   Hallway: 6 square centimeters

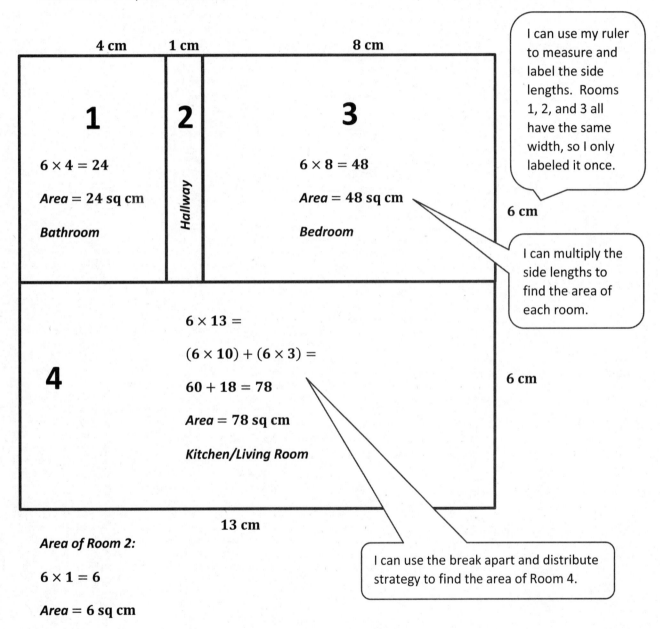

4 cm   **1 cm**   **8 cm**

1

$6 \times 4 = 24$

Area = 24 sq cm

Bathroom

2

Hallway

3

$6 \times 8 = 48$

Area = 48 sq cm

Bedroom

6 cm

4

$6 \times 13 =$

$(6 \times 10) + (6 \times 3) =$

$60 + 18 = 78$

Area = 78 sq cm

Kitchen/Living Room

6 cm

13 cm

> I can use my ruler to measure and label the side lengths. Rooms 1, 2, and 3 all have the same width, so I only labeled it once.

> I can multiply the side lengths to find the area of each room.

> I can use the break apart and distribute strategy to find the area of Room 4.

Area of Room 2:

$6 \times 1 = 6$

Area = 6 sq cm

Lesson 15:   Apply knowledge of area to determine areas of rooms in a given floor plan.

©2018 Great Minds®. eureka-math.org

319

Name _____   Date _____

Use a ruler to measure the side lengths of each numbered room in centimeters. Then, find the area. Use the measurements below to match, and label the rooms with the correct areas.

Kitchen: 45 square centimeters

Porch: 34 square centimeters

Bathroom: 24 square centimeters

Living Room: 63 square centimeters

Bedroom: 56 square centimeters

Hallway: 12 square centimeters

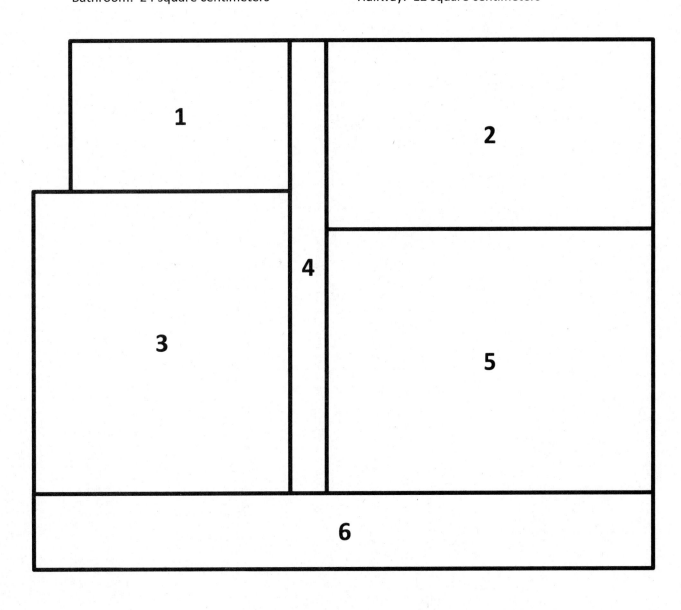

Lesson 15:   Apply knowledge of area to determine areas of rooms in a given floor plan.

©2018 Great Minds®. eureka-math.org

321

EUREKA
MATH™

Mrs. Harris designs her dream classroom on grid paper. The chart shows how much space she gives for each rectangular area. Use the information in the chart to draw and label a possible design for Mrs. Harris's classroom.

Reading area	48 square units	6 × 8
Carpet area	72 square units	9 × 8
Student desk area	90 square units	10 × 9
Science area	56 square units	7 × 8
Math area	64 square units	8 × 8

I can think of multiplication facts that equal each area. Then I can use the multiplication facts as the side lengths of each rectangular area. I can use the grid to help me draw each rectangular area.

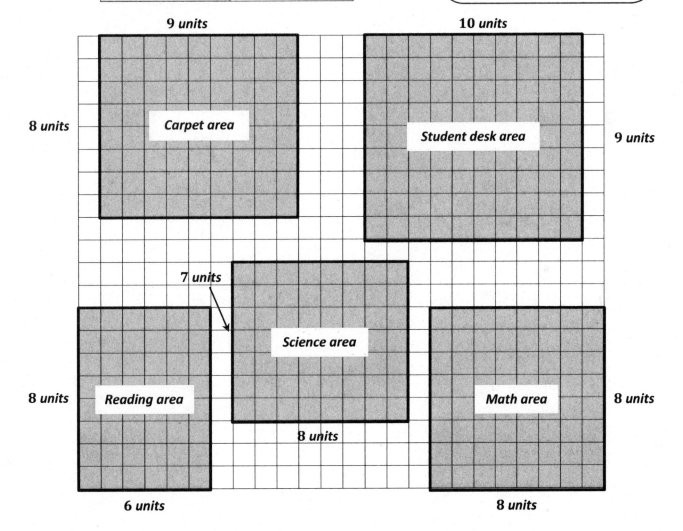

EUREKA MATH

Lesson 16:   Apply knowledge of area to determine areas of rooms in a given floor plan.

©2018 Great Minds®. eureka-math.org

323

Name _____ Date _____

Jeremy plans and designs his own dream playground on grid paper. His new playground will cover a total area of 100 square units. The chart shows how much space he gives for each piece of equipment, or area. Use the information in the chart to draw and label a possible way Jeremy can plan his playground.

Basketball court	10 square units
Jungle gym	9 square units
Slide	6 square units
Soccer area	24 square units

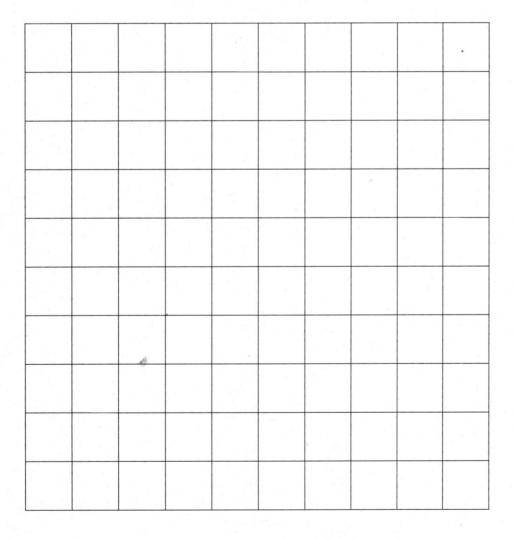

Lesson 16: Apply knowledge of area to determine areas of rooms in a given floor plan.

©2018 Great Minds®. eureka-math.org

325

EUREKA
MATH™

Credits

Great Minds® has made every effort to obtain permission for the reprinting of all copyrighted material. If any owner of copyrighted material is not acknowledged herein, please contact Great Minds for proper acknowledgment in all future editions and reprints of this module.